Atlas of AI

Atlas of AI

*Power, Politics, and the Planetary Costs
of Artificial Intelligence*

KATE CRAWFORD

Yale UNIVERSITY PRESS

New Haven and London

Yale University Press books may be purchased in quantity for
educational, business, or promotional use. For information,
please e-mail sales.press@yale.edu (U.S. office) or sales@yaleup
.co.uk (U.K. office).

Cover design and chapter opening illustrations by Vladan Joler.
Set in Minion by Tseng Information Systems, Inc.
Printed in the United States of America.

Library of Congress Control Number: 2020947842
ISBN 978-0-300-20957-0 (hardcover : alk. paper)
ISBN 978-0-300-26463-0 (paperback)

A catalogue record for this book is available from the British
Library.

10 9 8 7 6 5 4 3 2 1

For Elliott and Margaret

Contents

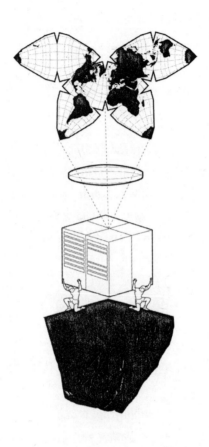

Introduction

The Smartest Horse in the World

At the end of the nineteenth century, Europe was captivated by a horse called Hans. "Clever Hans" was nothing less than a marvel: he could solve math problems, tell time, identify days on a calendar, differentiate musical tones, and spell out words and sentences. People flocked to watch the German stallion tap out answers to complex problems with his hoof and consistently arrive at the right answer. "What is two plus three?" Hans would diligently tap his hoof on the ground five times. "What day of the week is it?" The horse would then tap his hoof to indicate each letter on a purpose-built letter board and spell out the correct answer. Hans even mastered more complex questions, such as, "I have a number in mind. I subtract nine and have three as a remainder. What is the number?" By 1904, Clever Hans was an international celebrity, with the *New York Times* championing him as "Berlin's Wonderful Horse; He Can Do Almost Everything but Talk."[1]

Hans's trainer, a retired math teacher named Wilhelm von Osten, had long been fascinated by animal intelligence.

Von Osten had tried and failed to teach kittens and bear cubs cardinal numbers, but it wasn't until he started working with his own horse that he had success. He first taught Hans to count by holding the animal's leg, showing him a number, and then tapping on the hoof the correct number of times. Soon Hans responded by accurately tapping out simple sums. Next von Osten introduced a chalkboard with the alphabet spelled out, so Hans could tap a number for each letter on the board. After two years of training, von Osten was astounded by the animal's strong grasp of advanced intellectual concepts. So he took Hans on the road as proof that animals could reason. Hans became the viral sensation of the belle époque.

But many people were skeptical, and the German board of education launched an investigative commission to test Von Osten's scientific claims. The Hans Commission was led by the psychologist and philosopher Carl Stumpf and his assistant Oskar Pfungst, and it included a circus manager, a retired schoolteacher, a zoologist, a veterinarian, and a cavalry officer. Yet after extensive questioning of Hans, both with his trainer present and without, the horse maintained his record of correct answers, and the commission could find no evidence of deception. As Pfungst later wrote, Hans performed in front of "thousands of spectators, horse-fanciers, trick-trainers of first rank, and not one of them during the course of many months' observations are able to discover any kind of regular signal" between the questioner and the horse.[2]

The commission found that the methods Hans had been taught were more like "teaching children in elementary schools" than animal training and were "worthy of scientific examination."[3] But Strumpf and Pfungst still had doubts. One finding in particular troubled them: when the questioner did not know the answer or was standing far away, Hans rarely gave the correct answer. This led Pfungst and Strumpf to con-

Wilhelm von Osten and Clever Hans

sider whether some sort of unintentional signal had been providing Hans with the answers.

As Pfungst would describe in his 1911 book, their intuition was right: the questioner's posture, breathing, and facial expression would subtly change around the moment Hans reached the right answer, prompting Hans to stop there.[4] Pfungst later tested this hypothesis on human subjects and confirmed his result. What fascinated him most about this discovery was that questioners were generally unaware that they were providing pointers to the horse. The solution to the Clever Hans riddle, Pfungst wrote, was the unconscious direction from the horse's questioners.[5] The horse was trained to produce the results his owner wanted to see, but audiences felt that this was not the extraordinary intelligence they had imagined.

The story of Clever Hans is compelling from many angles: the relationship between desire, illusion, and action, the business of spectacles, how we anthropomorphize the nonhuman, how biases emerge, and the politics of intelligence. Hans inspired a term in psychology for a particular type of conceptual trap, the Clever Hans Effect or observer-expectancy effect, to describe the influence of experimenters' unintentional cues on their subjects. The relationship between Hans and von Osten points to the complex mechanisms by which biases find their ways into systems and how people become entangled with the phenomena they study. The story of Hans is now used in machine learning as a cautionary reminder that you can't always be sure of what a model has learned from the data it has been given.[6] Even a system that appears to perform spectacularly in training can make terrible predictions when presented with novel data in the world.

This opens a central question of this book: How is intelligence "made," and what traps can that create? At first glance, the story of Clever Hans is a story of how one man constructed intelligence by training a horse to follow cues and emulate humanlike cognition. But at another level, we see that the practice of making intelligence was considerably broader. The endeavor required validation from multiple institutions, including academia, schools, science, the public, and the military. Then there was the market for von Osten and his remarkable horse—emotional and economic investments that drove the tours, the newspaper stories, and the lectures. Bureaucratic authorities were assembled to measure and test the horse's abilities. A constellation of financial, cultural, and scientific interests had a part to play in the construction of Hans's intelligence and a stake in whether it was truly remarkable.

We can see two distinct mythologies at work. The first myth is that nonhuman systems (be it computers or horses) are analogues for human minds. This perspective assumes that with sufficient training, or enough resources, humanlike intelligence can be created from scratch, without addressing the fundamental ways in which humans are embodied, relational, and set within wider ecologies. The second myth is that intelligence is something that exists independently, as though it were natural and distinct from social, cultural, historical, and political forces. In fact, the concept of intelligence has done inordinate harm over centuries and has been used to justify relations of domination from slavery to eugenics.[7]

These mythologies are particularly strong in the field of artificial intelligence, where the belief that human intelligence can be formalized and reproduced by machines has been axiomatic since the mid-twentieth century. Just as Hans's intelligence was considered to be like that of a human, fostered carefully like a child in elementary school, so AI systems have repeatedly been described as simple but humanlike forms of intelligence. In 1950, Alan Turing predicted that "at the end of the century the use of words and general educated opinion will have altered so much that one will be able to speak of machines thinking without expecting to be contradicted."[8] The mathematician John von Neumann claimed in 1958 that the human nervous system is "prima facie digital."[9] MIT professor Marvin Minsky once responded to the question of whether machines could think by saying, "Of course machines can think; we can think and we are 'meat machines.'"[10] But not everyone was convinced. Joseph Weizenbaum, early AI inventor and creator of the first chatbot program, known as ELIZA, believed that the idea of humans as mere information processing systems is far too simplistic a notion of intelligence and that it drove the "perverse grand fantasy" that AI scientists could create a machine that learns "as a child does."[11]

This has been one of the core disputes in the history of artificial intelligence. In 1961, MIT hosted a landmark lecture series titled "Management and the Computer of the Future." A stellar lineup of computer scientists participated, including

Grace Hopper, J. C. R. Licklider, Marvin Minsky, Allen Newell, Herbert Simon, and Norbert Wiener, to discuss the rapid advances being made in digital computing. At its conclusion, John McCarthy boldly argued that the differences between human and machine tasks were illusory. There were simply some complicated human tasks that would take more time to be formalized and solved by machines.[12]

But philosophy professor Hubert Dreyfus argued back, concerned that the assembled engineers "do not even consider the possibility that the brain might process information in an entirely different way than a computer."[13] In his later work *What Computers Can't Do,* Dreyfus pointed out that human intelligence and expertise rely heavily on many unconscious and subconscious processes, while computers require all processes and data to be explicit and formalized.[14] As a result, less formal aspects of intelligence must be abstracted, eliminated, or approximated for computers, leaving them unable to process information about situations as humans do.

Much in AI has changed since the 1960s, including a shift from symbolic systems to the more recent wave of hype about machine learning techniques. In many ways, the early fights over what AI can do have been forgotten and the skepticism has melted away. Since the mid-2000s, AI has rapidly expanded as a field in academia and as an industry. Now a small number of powerful technology corporations deploy AI systems at a planetary scale, and their systems are once again hailed as comparable or even superior to human intelligence.

Yet the story of Clever Hans also reminds us how narrowly we consider or recognize intelligence. Hans was taught to mimic tasks within a very constrained range: add, subtract, and spell words. This reflects a limited perspective of what horses or humans can do. Hans was already performing remarkable feats of interspecies communication, public performance, and considerable patience, yet these were not recognized as intelligence. As author and engineer Ellen Ullman puts it, this belief that the mind is like a computer, and vice versa, has "infected decades of thinking in the computer and cognitive sciences," creating a kind of original sin for the field.[15] It is the ideology of Cartesian dualism in artificial intelligence: where AI is narrowly understood as disembodied intelligence, removed from any relation to the material world.

What Is AI? Neither Artificial nor Intelligent

Let's ask the deceptively simple question, What is artificial intelligence? If you ask someone in the street, they might mention Apple's Siri, Amazon's cloud service, Tesla's cars, or Google's search algorithm. If you ask experts in deep learning, they might give you a technical response about how neural nets are organized into dozens of layers that receive labeled data, are assigned weights and thresholds, and can classify data in ways that cannot yet be fully explained.[16] In 1978, when discussing expert systems, Professor Donald Michie described AI as knowledge refining, where "a reliability and competence of codification can be produced which far surpasses the highest level that the unaided human expert has ever, perhaps even could ever, attain."[17] In one of the most popular textbooks on the subject, Stuart Russell and Peter Norvig state that AI is the attempt to understand and build intelligent entities. "Intelligence is concerned mainly with rational action," they claim. "Ideally, an intelligent agent takes the best possible action in a situation."[18]

Each way of defining artificial intelligence is doing work, setting a frame for how it will be understood, measured, valued, and governed. If AI is defined by consumer brands for corporate infrastructure, then marketing and advertising have

predetermined the horizon. If AI systems are seen as more reliable or rational than any human expert, able to take the "best
possible action," then it suggests that they should be trusted to
make high-stakes decisions in health, education, and criminal justice. When specific algorithmic techniques are the sole
focus, it suggests that only continual technical progress matters, with no consideration of the computational cost of those
approaches and their far-reaching impacts on a planet under
strain.

In contrast, in this book I argue that AI is neither *artificial* nor *intelligent*. Rather, artificial intelligence is both
embodied and material, made from natural resources, fuel,
human labor, infrastructures, logistics, histories, and classifications. AI systems are not autonomous, rational, or able to
discern anything without extensive, computationally intensive
training with large datasets or predefined rules and rewards. In
fact, artificial intelligence as we know it depends entirely on a
much wider set of political and social structures. And due to
the capital required to build AI at scale and the ways of seeing
that it optimizes AI systems are ultimately designed to serve
existing dominant interests. In this sense, artificial intelligence
is a registry of power.

In this book we'll explore how artificial intelligence is
made, in the widest sense, and the economic, political, cultural, and historical forces that shape it. Once we connect AI
within these broader structures and social systems, we can escape the notion that artificial intelligence is a purely technical domain. At a fundamental level, AI is technical and social
practices, institutions and infrastructures, politics and culture.
Computational reason and embodied work are deeply inter-
linked: AI systems both reflect and produce social relations
and understandings of the world.

It's worth noting that the term "artificial intelligence"

can create discomfort in the computer science community. The phrase has moved in and out of fashion over the decades and is used more in marketing than by researchers. "Machine learning" is more commonly used in the technical literature. Yet the nomenclature of AI is often embraced during funding application season, when venture capitalists come bearing checkbooks, or when researchers are seeking press attention for a new scientific result. As a result, the term is both used and rejected in ways that keep its meaning in flux. For my purposes, I use AI to talk about the massive industrial formation that includes politics, labor, culture, and capital. When I refer to machine learning, I'm speaking of a range of technical approaches (which are, in fact, social and infrastructural as well, although rarely spoken about as such).

But there are significant reasons *why* the field has been focused so much on the technical—algorithmic breakthroughs, incremental product improvements, and greater convenience. The structures of power at the intersection of technology, capital, and governance are well served by this narrow, abstracted analysis. To understand how AI is fundamentally political, we need to go beyond neural nets and statistical pattern recognition to instead ask *what* is being optimized, and *for whom*, and *who* gets to decide. Then we can trace the implications of those choices.

Seeing AI Like an Atlas

How can an atlas help us to understand how artificial intelligence is made? An atlas is an unusual type of book. It is a collection of disparate parts, with maps that vary in resolution from a satellite view of the planet to a zoomed-in detail of an archipelago. When you open an atlas, you may be seeking specific information about a particular place—or perhaps you are wandering, following your curiosity, and finding unexpected pathways and new perspectives. As historian of science Lorraine Daston observes, all scientific atlases seek to school the eye, to focus the observer's attention on particular telling details and significant characteristics.[19] An atlas presents you with a particular viewpoint of the world, with the imprimatur of science—scales and ratios, latitudes and longitudes—and a sense of form and consistency.

Yet an atlas is as much an act of creativity—a subjective, political, and aesthetic intervention—as it is a scientific collection. The French philosopher Georges Didi-Huberman thinks of the atlas as something that inhabits the aesthetic paradigm of the visual and the epistemic paradigm of knowledge. By implicating both, it undermines the idea that science and art are ever completely separate.[20] Instead, an atlas offers us the possibility of rereading the world, linking disparate pieces differently and "reediting and piecing it together again without thinking we are summarizing or exhausting it."[21]

Perhaps my favorite account of how a cartographic approach can be helpful comes from the physicist and technology critic Ursula Franklin: "Maps represent purposeful endeavors: they are meant to be useful, to assist the traveler and bridge the gap between the known and the as yet unknown; they are testaments of collective knowledge and insight."[22]

Maps, at their best, offer us a compendium of open pathways—shared ways of knowing—that can be mixed and combined to make new interconnections. But there are also maps of domination, those national maps where territory is carved along the fault lines of power: from the direct interventions of drawing borders across contested spaces to revealing the colonial paths of empires. By invoking an atlas, I'm suggesting that we need new ways to understand the empires of artificial intelligence. We need a theory of AI that accounts for the states and corporations that drive and dominate it, the extractive min-ing that leaves an imprint on the planet, the mass capture of data, and the profoundly unequal and increasingly exploitative labor practices that sustain it. These are the shifting tecton-ics of power in AI. A topographical approach offers different perspectives and scales, beyond the abstract promises of arti-ficial intelligence or the latest machine learning models. The aim is to understand AI in a wider context by walking through the many different landscapes of computation and seeing how they connect.[23]

There's another way in which atlases are relevant here. The field of AI is explicitly attempting to capture the planet in a computationally legible form. This is not a metaphor so much as the industry's direct ambition. The AI industry is making and normalizing its own proprietary maps, as a cen-tralized God's-eye view of human movement, communication, and labor. Some AI scientists have stated their desire to cap-ture the world and to supersede other forms of knowing. AI professor Fei-Fei Li describes her ImageNet project as aiming to "map out the entire world of objects."[24] In their textbook, Russell and Norvig describe artificial intelligence as "relevant to any intellectual task; it is truly a universal field."[25] One of the founders of artificial intelligence and early experimenter in facial recognition, Woody Bledsoe, put it most bluntly: "in the long run, AI is the *only* science."[26] This is a desire not to create an atlas of the world but to be *the* atlas—the dominant way of seeing. This colonizing impulse centralizes power in the AI field: it determines how the world is measured and de-fined while simultaneously denying that this is an inherently political activity.

Instead of claiming universality, this book is a partial ac-count, and by bringing you along on my investigations, I hope to show you how my views were formed. We will encounter well-visited and lesser-known landscapes of computation: the pits of mines, the long corridors of energy-devouring data centers, skull archives, image databases, and the fluorescent-lit hangars of delivery warehouses. These sites are included not just to illustrate the material construction of AI and its ideologies but also to "illuminate the unavoidably subjective and political aspects of mapping, and to provide alternatives to hegemonic, authoritative—and often naturalized and reified—approaches," as media scholar Shannon Mattern writes.[27]

Models for understanding and holding systems accountable have long rested on ideals of transparency. As I've written with the media scholar Mike Ananny, being able to *see* a system is sometimes equated with being able to know how it works and how to govern it.[28] But this tendency has serious limitations. In the case of AI, there is no singular black box to open, no secret to expose, but a multitude of interlaced systems of power. Complete transparency, then, is an impossible goal. Rather, we gain a better understanding of AI's role in the world by engaging with its material architectures, contextual environments, and prevailing politics and by tracing how they are connected.

My thinking in this book has been informed by the disciplines of science and technology studies, law, and political philosophy and from my experience working in both academia and an industrial AI research lab for almost a decade. Over those years, many generous colleagues and communities have changed the way I see the world: mapping is always a collective exercise, and this is no exception.[29] I'm grateful to the scholars who created new ways to understand sociotechnical systems, including Geoffrey Bowker, Benjamin Bratton, Wendy Chun, Lorraine Daston, Peter Galison, Ian Hacking, Stuart Hall, Donald MacKenzie, Achille Mbembé, Alondra Nelson, Susan Leigh Star, and Lucy Suchman, among many others. This book benefited from many in-person conversations and reading the recent work by authors studying the politics of technology, including Mark Andrejevic, Ruha Benjamin, Meredith Broussard, Simone Browne, Julie Cohen, Sasha Costanza-Chock, Virginia Eubanks, Tarleton Gillespie, Mar Hicks, Tung-Hui Hu, Yuk Hui, Safiya Umoja Noble, and Astra Taylor.

As with any book, this one emerges from a specific lived experience that imposes limitations. As someone who has lived and worked in the United States for the past decade, my focus skews toward the AI industry in Western centers of power. But my aim is not to create a complete global atlas—the very idea invokes capture and colonial control. Instead, any author's view can be only partial, based on local observations and interpretations, in what environmental geographer Samantha Saville calls a "humble geography" that acknowledges one's specific perspectives rather than claiming objectivity or mastery.[30]

Just as there are many ways to make an atlas, so there are many possible futures for how AI will be used in the world. The expanding reach of AI systems may seem inevitable, but this is contestable and incomplete. The underlying visions of the AI field do not come into being autonomously but instead have been constructed from a particular set of beliefs and perspectives. The chief designers of the contemporary atlas of AI are a small and homogenous group of people, based in a handful of cities, working in an industry that is currently the wealthiest in the world. Like medieval European *mappae mundi,* which illustrated religious and classical concepts as much as coordinates, the maps made by the AI industry are political interventions, as opposed to neutral reflections of the world. This book is made against the spirit of colonial mapping logics, and it embraces different stories, locations, and knowledge bases to better understand the role of AI in the world.

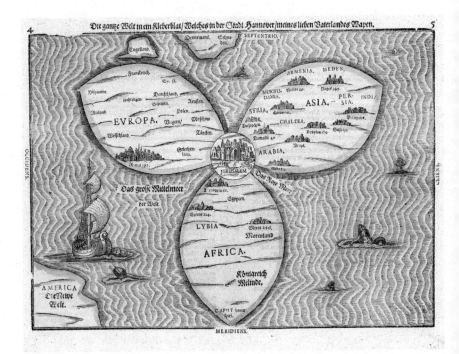

Heinrich Bünting's *mappa mundi*, known as *The Bünting Clover Leaf Map*, which symbolizes the Christian Trinity, with the city of Jerusalem at the center of the world. From *Itinerarium Sacrae Scripturae* (Magdeburg, 1581)

Topographies of Computation

How, at this moment in the twenty-first century, is AI conceptualized and constructed? What is at stake in the turn to artificial intelligence, and what kinds of politics are contained in the way these systems map and interpret the world? What are the social and material consequences of including AI and related algorithmic systems into the decision-making systems of social institutions like education and health care, finance, government operations, workplace interactions and hiring, communication systems, and the justice system? This book is not a story about code and algorithms or the latest thinking in computer vision or natural language processing or reinforcement learning. Many other books do that. Neither is it an ethnographic account of a single community and the effects of AI on their experience of work or housing or medicine—although we certainly need more of those.

Instead, this is an expanded view of artificial intelligence as an *extractive industry*. The creation of contemporary AI systems depends on exploiting energy and mineral resources from the planet, cheap labor, and data at scale. To observe this in action, we will go on a series of journeys to places that reveal the makings of AI.

In chapter 1, we begin in the lithium mines of Nevada, one of the many sites of mineral extraction needed to power contemporary computation. Mining is where we see the extractive politics of AI at their most literal. The tech sector's demand for rare earth minerals, oil, and coal is vast, but the true costs of this extraction is never borne by the industry itself. On the software side, building models for natural language processing and computer vision is enormously energy hungry, and the competition to produce faster and more efficient models has driven computationally greedy methods that expand AI's carbon footprint. From the last trees in Malaysia that were harvested to produce latex for the first transatlantic undersea cables to the giant artificial lake of toxic residues in Inner Mongolia, we trace the environmental and human birthplaces of planetary computation networks and see how they continue to terraform the planet.

Chapter 2 shows how artificial intelligence is made of human labor. We look at the digital pieceworkers paid pennies on the dollar clicking on microtasks so that data systems can seem more intelligent than they are.[31] Our journey will take us inside the Amazon warehouses where employees must keep in time with the algorithmic cadences of a vast logistical empire, and we will visit the Chicago meat laborers on the disassembly lines where animal carcasses are vivisected and prepared for consumption. And we'll hear from the workers who are protesting against the way that AI systems are increasing surveillance and control for their bosses.

Labor is also a story about time. Coordinating the actions of humans with the repetitive motions of robots and line machinery has always involved a controlling of bodies in space and time.[32] From the invention of the stopwatch to Google's TrueTime, the process of time coordination is at the heart of workplace management. AI technologies both require and create the conditions for ever more granular and precise mechanisms of temporal management. Coordinating time demands increasingly detailed information about what people are doing and how and when they do it.

Chapter 3 focuses on the role of data. All publicly accessible digital material—including data that is personal or potentially damaging—is open to being harvested for training datasets that are used to produce AI models. There are gigantic datasets full of people's selfies, of hand gestures, of people driving cars, of babies crying, of newsgroup conversations from the 1990s, all to improve algorithms that perform such functions as facial recognition, language prediction, and object detection. When these collections of data are no longer seen as people's personal material but merely as *infrastructure,* the specific meaning or context of an image or a video is assumed to be irrelevant. Beyond the serious issues of privacy and ongoing surveillance capitalism, the current practices of working with data in AI raise profound ethical, methodological, and epistemological concerns.[33]

And how is all this data used? In chapter 4, we look at

the practices of classification in artificial intelligence systems, what sociologist Karin Knorr Cetina calls the "epistemic machinery."[34] We see how contemporary systems use labels to predict human identity, commonly using binary gender, essentialized racial categories, and problematic assessments of character and credit worthiness. A sign will stand in for a system, a proxy will stand for the real, and a toy model will be asked to substitute for the infinite complexity of human subjectivity. By looking at how classifications are made, we see how technical schemas enforce hierarchies and magnify inequity. Machine learning presents us with a regime of normative reasoning that, when in the ascendant, takes shape as a powerful governing rationality.

From here, we travel to the hill towns of Papua New Guinea to explore the history of affect recognition, the idea that facial expressions hold the key to revealing a person's inner emotional state. Chapter 5 considers the claim of the psychologist Paul Ekman that there are a small set of universal emotional states which can be read directly from the face. Tech companies are now deploying this idea in affect recognition systems, as part of an industry predicted to be worth more than seventeen billion dollars.[35] But there is considerable scientific controversy around emotion detection, which is at best incomplete and at worst misleading. Despite the unstable premise, these tools are being rapidly implemented into hiring, education, and policing systems.

In chapter 6 we look at the ways in which AI systems are used as a tool of state power. The military past and present of artificial intelligence have shaped the practices of surveillance, data extraction, and risk assessment we see today. The deep interconnections between the tech sector and the military are now being reined in to fit a strong nationalist agenda. Meanwhile, extralegal tools used by the intelligence community have now dispersed, moving from the military world into the commercial technology sector, to be used in classrooms, police stations, workplaces, and unemployment offices. The military logics that have shaped AI systems are now part of the workings of municipal government, and they are further skewing the relation between states and subjects.

The concluding chapter assesses how artificial intelligence functions as a structure of power that combines infrastructure, capital, and labor. From the Uber driver being nudged to the undocumented immigrant being tracked to the public housing tenants contending with facial recognition systems in their homes, AI systems are built with the logics of capital, policing, and militarization—and this combination further widens the existing asymmetries of power. These ways of seeing depend on the twin moves of abstraction and extraction: abstracting away the material conditions of their making while extracting more information and resources from those least able to resist.

But these logics can be challenged, just as systems that perpetuate oppression can be rejected. As conditions on Earth change, calls for data protection, labor rights, climate justice, and racial equity should be heard together. When these interconnected movements for justice inform how we understand artificial intelligence, different conceptions of planetary politics become possible.

Extraction, Power, and Politics

Artificial intelligence, then, is an idea, an infrastructure, an industry, a form of exercising power, and a way of seeing; it's also a manifestation of highly organized capital backed by vast systems of extraction and logistics, with supply chains that wrap around the entire planet. All these things are part of what artificial intelligence is—a two-word phrase onto which is mapped a complex set of expectations, ideologies, desires, and fears.

AI can seem like a spectral force—as disembodied computation—but these systems are anything but abstract. They are physical infrastructures that are reshaping the Earth, while simultaneously shifting how the world is seen and understood.

It's important for us to contend with these many aspects of artificial intelligence—its malleability, its messiness, and its spatial and temporal reach. The promiscuity of AI as a term, its openness to being reconfigured, also means that it can be put to use in a range of ways: it can refer to everything from consumer devices like the Amazon Echo to nameless back-end processing systems, from narrow technical papers to the biggest industrial companies in the world. But this has its usefulness, too. The breadth of the term "artificial intelligence" gives us license to consider all these elements and how they are deeply imbricated: from the politics of intelligence to the mass harvesting of data; from the industrial concentration of the tech sector to geopolitical military power; from the deracinated environment to ongoing forms of discrimination.

The task is to remain sensitive to the terrain and to watch the shifting and plastic meanings of the term "artificial intelligence"—like a container into which various things are placed and then removed—because that, too, is part of the story.

Simply put, artificial intelligence is now a player in the shaping of knowledge, communication, and power. These reconfigurations are occurring at the level of epistemology, principles of justice, social organization, political expression, culture, understandings of human bodies, subjectivities, and identities: what we are and what we can be. But we can go further. Artificial intelligence, in the process of remapping and intervening in the world, is politics by other means—although rarely acknowledged as such. These politics are driven by the Great Houses of AI, which consist of the half-dozen or so companies that dominate large-scale planetary computation.

Many social institutions are now influenced by these tools and methods, which shape what they value and how decisions are made while creating a complex series of downstream effects. The intensification of technocratic power has been under way for a long time, but the process has now accelerated. In part this is due to the concentration of industrial capital at a time of economic austerity and outsourcing, including the defunding of social welfare systems and institutions that once acted as a check on market power. This is why we must contend with AI as a political, economic, cultural, and scientific force. As Alondra Nelson, Thuy Linh Tu, and Alicia Headlam Hines observe, "Contests around technology are always linked to larger struggles for economic mobility, political maneuvering, and community building."[36]

We are at a critical juncture, one that requires us to ask hard questions about the way AI is produced and adopted. We need to ask: What is AI? What forms of politics does it propagate? Whose interests does it serve, and who bears the greatest risk of harm? And where should the use of AI be constrained? These questions will not have easy answers. But neither is this an irresolvable situation or a point of no return—dystopian forms of thinking can paralyze us from taking action and prevent urgently needed interventions.[37] As Ursula Franklin writes, "The viability of technology, like democracy, depends in the end on the practice of justice and on the enforcement of limits to power."[38]

This book argues that addressing the foundational problems of AI and planetary computation requires connecting issues of power and justice: from epistemology to labor rights, resource extraction to data protections, racial inequity to climate change. To do that, we need to expand our understanding of what is under way in the empires of AI, to see what is at stake, and to make better collective decisions about what should come next.

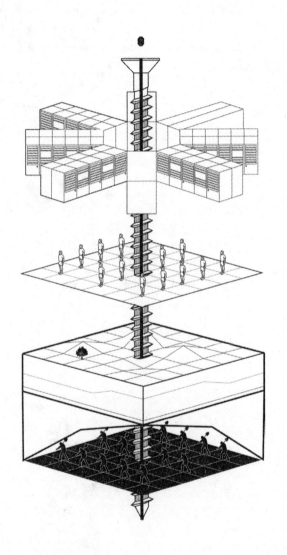

1
Earth

The Boeing 757 banks right over San Jose on its final approach to San Francisco International Airport. The left wing drops as the plane lines up with the runway, revealing an aerial view of the tech sector's most iconic location. Below are the great empires of Silicon Valley. The gigantic black circle of Apple's headquarters is laid out like an uncapped camera lens, glistening in the sun. Then there's Google's head office, nestled close to NASA's Moffett Federal Airfield. This was once a key site for the U.S. Navy during World War II and the Korean War, but now Google has a sixty-year lease on it, and senior executives park their private planes here. Arrayed near Google are the large manufacturing sheds of Lockheed Martin, where the aerospace and weapons manufacturing company builds hundreds of orbital satellites destined to look down on the activities of Earth. Next, by the Dumbarton Bridge, appears a collection of squat buildings that are home to Facebook, ringed with massive parking lots close to the sulfuric salt ponds of the Ravenswood Slough. From this vantage point, the nondescript suburban cul-de-sacs and industrial midrise skyline of Palo Alto betray little of its true wealth, power, and influence. There are only a few hints of its centrality in the global economy and in the computational infrastructure of the planet.

I'm here to learn about artificial intelligence and what it is made from. To see that, I will need to leave Silicon Valley altogether.

From the airport, I jump into a van and drive east. I cross the San Mateo–Hayward Bridge and pass by the Lawrence Livermore National Laboratory, where Edward Teller directed his research into thermonuclear weapons in the years after World War II. Soon the Sierra Nevada foothills rise beyond the Central Valley towns of Stockton and Manteca. Here the roads start winding up through the tall granite cliffs of the Sonora Pass and down the eastern side of the mountains toward grassy valleys dotted with golden poppies. Pine forests give way to the alkaline waters of Mono Lake and the parched desert landforms of the Basin and Range. To refuel, I pull into Hawthorne, Nevada, site of the world's biggest ammunition depot, where the U.S. Army stores armaments in dozens of dirt-covered ziggurats that populate the valley in neat rows. Driving along Nevada State Route 265 I see a lone VORTAC in the distance, a large bowling pin–shaped radio tower that was designed for the era before GPS. It has a single function: it broadcasts "I am here" to all passing aircraft, a fixed point of reference in a lonely terrain.

My destination is the unincorporated community of Silver Peak in Nevada's Clayton Valley, where about 125 people live, depending on how you count. The mining town, one of the oldest in Nevada, was almost abandoned in 1917 after the ground was stripped bare of silver and gold. A few gold rush buildings still stand, eroding under the desert sun. The town may be small, with more junked cars than people, but it harbors something exceedingly rare. Silver Peak is perched on the edge of a massive underground lake of lithium. The valuable

Silver Peak Lithium Mine. Photograph by Kate Crawford

lithium brine under the surface is pumped out of the ground and left in open, iridescent green ponds to evaporate. From miles away, the ponds can be seen when they catch the light and shimmer. Up close, it's a different view. Alien-looking black pipes erupt from the ground and snake along the salt-encrusted earth, moving in and out of shallow trenches, ferrying the salty cocktail to its drying pans.

Here, in a remote pocket of Nevada, is a place where the stuff of AI is made.

Mining for AI

Clayton Valley is connected to Silicon Valley in much the way that the nineteenth-century goldfields were to early San Francisco. The history of mining, like the devastation it leaves in its wake, is commonly overlooked in the strategic amnesia that accompanies stories of technological progress. As historical geographer Gray Brechin points out, San Francisco was built from the gains of pulling gold and silver out of the lands of California and Nevada in the 1800s.[1] The city is made from mining. Those same lands had been taken from Mexico under the Treaty of Guadalupe Hidalgo in 1848 at the end of the Mexican-American War, when it was already clear to the settlers that these would be highly valuable goldfields. It was a textbook example, Brechin observes, of the old adage that "commerce follows the flag, but the flag follows the pick."[2] Thousands of people were forced from their homes during this substantial territorial expansion of the United States. After America's imperial invasion, the miners moved in. The land was stripped until the waterways were contaminated and the surrounding forests destroyed.

Since antiquity, the business of mining has only been profitable because it does not have to account for its true costs: including environmental damage, the illness and death of miners, and the loss to the communities it displaces. In 1555, Georgius Agricola, known as the father of mineralogy, observed that "it is clear to all that there is greater detriment from mining than the value of the metals which the mining produces."[3] In other words, those who profit from mining do so only because the costs must be sustained by others, those living and those not yet born. It is easy to put a price on precious metals, but what is the exact value of a wilderness, a clean stream, breathable air, the health of local communities? It was never estimated, and thus an easy calculus emerged: extract everything as rapidly as possible. It was the "move fast and break things" of a different time. The result was that the Central Valley was decimated, and as one tourist observed in 1869,

"Tornado, flood, earthquake and volcano combined could hardly make greater havoc, spread wider ruin and wreck than [the] gold-washing operations. . . . There are no rights which mining respects in California. It is the one supreme interest."[4]

As San Francisco drew enormous wealth from the mines, it was easy for its populace to forget where it all came from. The mines were located far from the city they enriched, and this remoteness allowed city dwellers to remain ignorant of what was happening to the mountains, rivers, and laborers that fed their fortunes. But small reminders of the mines are all around. The city's new buildings used the same technology that came from deep within the Central Valley for transport and life support. The pulley systems that carried miners down into the mine shafts were adapted and turned upside down to transport people in elevators to the top of the city's high-rises.[5] Brechin suggests that we should think of the skyscrapers of San Francisco as inverted minescapes. The ores extracted from holes in the ground were sold to create the stories in the air; the deeper the extractions went, the higher the great towers of office work stretched into the sky.

San Francisco is enriched once more. Once it was gold ore that underwrote fortunes; now it is the extraction of substances like white lithium crystal. It's known in mineral markets as "gray gold."[6] The technology industry has become a new supreme interest, and the five biggest companies in the world by market capitalization have offices in this city: Apple, Microsoft, Amazon, Facebook, and Google. Walking past the start-up warehouses in the SoMa district where miners in tents once lived, you can see luxury cars, venture capital–backed coffee chains, and sumptuous buses with tinted windows running along private routes, carrying workers to their offices in Mountain View or Menlo Park.[7] But only a short walk away is Division Street, a multilane thoroughfare between SoMa and the Mission district, where rows of tents have returned to shelter people who have nowhere to go. In the wake of the tech boom, San Francisco now has one of the highest rates of street homelessness in the United States.[8] The United Nations special rapporteur on adequate housing called it an "unacceptable" human rights violation, due to the thousands of homeless residents denied basic necessities of water, sanitation, and health services in contrast to the record number of billionaires who live nearby.[9] The greatest benefits of extraction have been captured by the few.

In this chapter we'll traverse across Nevada, San Jose, and San Francisco, as well as Indonesia, Malaysia, China, and Mongolia: from deserts to oceans. We'll also walk the spans of historical time, from conflict in the Congo and artificial black lakes in the present day to the Victorian passion for white latex. The scale will shift, telescoping from rocks to cities, trees to megacorporations, transoceanic shipping lanes to the atomic bomb. But across this planetary supersystem we will see the logics of extraction, a constant drawdown of minerals, water, and fossil fuels, undergirded by the violence of wars, pollution, extinction, and depletion. The effects of large-scale computation can be found in the atmosphere, the oceans, the earth's crust, the deep time of the planet, and the brutal impacts on disadvantaged populations around the world. To understand it all, we need a panoramic view of the planetary scale of computational extraction.

Landscapes of Computation

I'm driving through the desert valley on a summer afternoon to see the workings of this latest mining boom. I ask my phone to direct me to the perimeter of the lithium ponds, and it replies from its awkward perch on the dashboard, tethered by a white USB cable. Silver Peak's large, dry lake bed was formed millions of years ago during the late Tertiary Period. It's surrounded by crusted stratifications pushing up into ridgelines containing dark limestones, green quartzites, and gray and red slate.[10] Lithium was discovered here after the area was scoped for strategic minerals like potash during World War II. This soft, silvery metal was mined in only modest quantities for the next fifty years, until it became highly valuable material for the technology sector.

In 2014, Rockwood Holdings, Inc., a lithium mining operation, was acquired by the chemical manufacturing company Albemarle Corporation for $6.2 billion. It is the only operating lithium mine in the United States. This makes Silver Peak a site of intense interest to Elon Musk and the many other tech tycoons for one reason: rechargeable batteries. Lithium is a crucial element for their production. Smartphone batteries, for example, usually contain about three-tenths of an ounce of it. Each Tesla Model S electric car needs about one hundred thirty-eight pounds of lithium for its battery pack.[11] These kinds of batteries were never intended to supply a machine as power hungry as a car, but lithium batteries are currently the only mass-market option available.[12] All of these batteries have a limited lifespan; once degraded, they are discarded as waste.

About two hundred miles north of Silver Peak is the Tesla Gigafactory. This is the world's largest lithium battery plant. Tesla is the number-one lithium-ion battery consumer in the world, purchasing them in high volumes from Panasonic and Samsung and repackaging them in its cars and home chargers. Tesla is estimated to use more than twenty-eight thousand tons of lithium hydroxide annually—half of the planet's total consumption.[13] In fact, Tesla could more accurately be described as a battery business than a car company.[14] The imminent shortage of such critical minerals as nickel, copper, and lithium poses a risk for the company, making the lithium lake at Silver Peak highly desirable.[15] Securing control of the mine would mean controlling the U.S. domestic supply.

As many have shown, the electric car is far from a perfect solution to carbon dioxide emissions.[16] The mining, smelting, export, assemblage, and transport of the battery supply chain has a significant negative impact on the environment and, in turn, on the communities affected by its degradation. A small number of home solar systems produce their own energy. But for the majority of cases, charging an electric car necessitates taking power from the grid, where currently less than a fifth of all electricity in the United States comes from renewable energy sources.[17] So far none of this has dampened the determination of auto manufacturers to compete with Tesla, putting increasing pressure on the battery market and accelerating the removal of diminishing stores of the necessary minerals.

Global computation and commerce rely on batteries. The term "artificial intelligence" may invoke ideas of algorithms, data, and cloud architectures, but none of that can function without the minerals and resources that build computing's core components. Rechargeable lithium-ion batteries are essential for mobile devices and laptops, in-home digital assistants, and data center backup power. They undergird the internet and every commerce platform that runs on it, from banking to retail to stock market trades. Many aspects of modern life have been moved to "the cloud" with little consideration of these material costs. Our work and personal lives, our medical histories, our leisure time, our entertainment, our political interests—all of this takes place in the world of networked computing architectures that we tap into from devices we hold in one hand, with lithium at their core.

The mining that makes AI is both literal and metaphorical. The new extractivism of data mining also encompasses and propels the old extractivism of traditional mining. The stack required to power artificial intelligence systems goes well beyond the multilayered technical stack of data modeling, hardware, servers, and networks. The full-stack supply chain of AI reaches into capital, labor, and Earth's resources—and from each, it demands an enormous amount.[18] The cloud is the backbone of the artificial intelligence industry, and it's made of rocks and lithium brine and crude oil.

In his book *A Geology of Media,* theorist Jussi Parikka suggests we think of media not from Marshall McLuhan's point of view—in which media are extensions of the human senses—but rather as extensions of Earth.[19] Computational media now participate in geological (and climatological) processes, from the transformation of the earth's materials into infrastructures and devices to the powering of these new systems with oil and gas reserves. Reflecting on media and technology as geological processes enables us to consider the radical depletion of nonrenewable resources required to drive the technologies of the present moment. Each object in the extended network of an AI system, from network routers to batteries to data centers, is built using elements that required billions of years to form inside the earth.

From the perspective of deep time, we are extracting Earth's geological history to serve a split second of contemporary technological time, building devices like the Amazon Echo and the iPhone that are often designed to last for only a few years. The Consumer Technology Association notes that the average smartphone life span is a mere 4.7 years.[20] This obsolescence cycle fuels the purchase of more devices, drives up profits, and increases incentives for the use of unsustainable extraction practices. After a slow process of development, these minerals, elements, and materials then go through an extraordinarily rapid period of excavation, processing, mixing, smelting, and logistical transport—crossing thousands of miles in their transformation. What begins as ore removed from the ground, after the spoil and the tailings are discarded, is then made into devices that are used and discarded. They ultimately end up buried in e-waste dumping grounds in places like Ghana and Pakistan. The lifecycle of an AI system from birth to death has many fractal supply chains: forms of exploitation of human labor and natural resources and massive concentrations of corporate and geopolitical power. And all along the chain, a continual, large-scale consumption of energy keeps the cycle going.

The extractivism on which San Francisco was built is echoed in the practices of the tech sector based there today.[21] The massive ecosystem of AI relies on many kinds of extraction: from harvesting the data made from our daily activities and expressions, to depleting natural resources, and to exploiting labor around the globe so that this vast planetary network can be built and maintained. And AI extracts far more from us and the planet than is widely known. The Bay Area is a central node in the mythos of AI, but we'll need to traverse far beyond the United States to see the many-layered legacies of human and environmental damage that have powered the tech industry.

The Mineralogical Layer

The lithium mines in Nevada are just one of the places where the materials are extracted from the earth's crust to make AI. There are many such sites, including the Salar in southwest Bolivia—the richest site of lithium in the world and thus a site of ongoing political tension—as well as places in central Congo, Mongolia, Indonesia, and the Western Australia deserts. These are the other birthplaces of AI in the greater geography of industrial extraction. Without the minerals from these locations, contemporary computation simply does not work. But these materials are in increasingly short supply.

In 2020, scientists at the U.S. Geological Survey published a short list of twenty-three minerals that are a high "supply risk" to manufacturers, meaning that if they became unavailable, entire industries—including the tech sector—would grind to a halt.[22] The critical minerals include the rare earth elements dysprosium and neodymium, which are used inside iPhone speakers and electric vehicle motors; germanium, which is used in infrared military devices for soldiers and in drones; and cobalt, which improves performance for lithium-ion batteries.

There are seventeen rare earth elements: lanthanum, cerium, praseodymium, neodymium, promethium, samarium, europium, gadolinium, terbium, dysprosium, holmium, erbium, thulium, ytterbium, lutetium, scandium, and yttrium. They are processed and embedded in laptops and smartphones, making those devices smaller and lighter. The elements can be found in color displays, speakers, camera lenses, rechargeable batteries, hard drives, and many other components. They are key elements in communication systems, from fiber-optic cables and signal amplification in mobile communication towers to satellites and GPS technology. But extracting these minerals from the ground often comes with local and geopolitical violence. Mining is and always has been a brutal undertaking. As Lewis Mumford writes, "Mining was the key industry that furnished the sinews of war and increased the metallic contents of the original capital hoard, the war chest: on the other hand, it furthered the industrialization of arms, and enriched the financier by both processes."[23] To understand the business of AI, we must reckon with the war, famine, and death that mining brings with it.

Recent U.S. legislation that regulates some of those seventeen rare earth elements only hints at the devastation associated with their extraction. The 2010 Dodd-Frank Act focused on reforming the financial sector in the wake of the 2008 financial crisis. It included a specific provision about so-called *conflict minerals,* or natural resources extracted in a conflict zone and then sold to fund the conflict. Companies using gold, tin, tungsten, and tantalum from the region around the Democratic Republic of the Congo now had a reporting requirement to track where those minerals came from and whether the sale was funding armed militia in the region.[24] Like "conflict diamonds," the term "conflict minerals" masks the profound suffering and prolific killing in the mining sector. Mining profits have financed military operations in the decades-long Congo-area conflict, fueling the deaths of thousands and the displacement of millions.[25] Furthermore, working conditions inside the mines have often amounted to modern slavery.[26]

It took Intel more than four years of sustained effort to develop basic insight into its own supply chain.[27] Intel's supply chain is complex, with more than sixteen thousand suppliers in over a hundred countries providing direct materials for the company's production processes, tools, and machines for their factories, as well as their logistics and packaging services.[28] In addition, Intel and Apple have been criticized for auditing only smelters—not the actual mines—to determine the conflict-free status of minerals. The tech giants were assessing smelting plants outside of Congo, and the audits were often performed by locals. So even the conflict-free certifications of the tech industry are now under question.[29]

Dutch-based technology company Philips has also claimed that it was working to make its supply chain "conflict-free." Like Intel, Philips has tens of thousands of suppliers, each of which provides component parts for the company's manufacturing processes.[30] Those suppliers are themselves linked downstream to thousands of component manufacturers acquiring treated materials from dozens of smelters. The smelters in turn buy their materials from an unknown number of traders who deal directly with both legal and illegal mining operations to source the various minerals that end up in computer components.[31]

According to the computer manufacturer Dell, the complexities of the metals and mineral supply chains pose almost insurmountable challenges to the production of conflict-free electronics components. The elements are laundered through such a vast number of entities along the chain that sourcing their provenance proves impossible—or so the end-product manufacturers claim, allowing them a measure of plausible deniability for any exploitative practices that drive their profits.[32]

Just like the mines that served San Francisco in the nineteenth century, extraction for the technology sector is done by keeping the real costs out of sight. Ignorance of the supply chain is baked into capitalism, from the way businesses protect themselves through third-party contractors and suppliers to the way goods are marketed and advertised to consumers. More than plausible deniability, it has become a well-practiced form of bad faith: the left hand cannot know what the right hand is doing, which requires increasingly lavish, baroque, and complex forms of distancing.

While mining to finance war is one of the most extreme cases of harmful extraction, most minerals are not sourced from direct war zones. This doesn't mean, however, that they are free from human suffering and environmental destruction. The focus on conflict minerals, though important, has also been used to avert focus from the harms of mining writ large.

If we visit the primary sites of mineral extraction for computational systems, we find the repressed stories of acid-bleached rivers and deracinated landscapes and the extinction of plant and animal species that were once vital to the local ecology.

Black Lakes and White Latex

In Baotou, the largest city in Inner Mongolia, there is an artificial lake filled with toxic black mud. It reeks of sulfur and stretches as far as the eye can see, covering more than five and a half miles in diameter. The black lake contains more than 180 million tons of waste powder from ore processing.[33] It was created by the waste runoff from the nearby Bayan Obo mines, which is estimated to contain almost 70 percent of the world's reserves of rare earth minerals. It is the largest deposit of rare earth elements on the planet.[34]

China supplies 95 percent of the world's rare earth minerals. China's market domination, as the writer Tim Maughan observes, owes far less to geology than to the country's willingness to take on the environmental damage of extraction.[35] Although rare earth minerals like neodymium and cerium are relatively common, making them usable requires the hazardous process of dissolving them in large volumes of sulfuric and nitric acid. These acid baths yield reservoirs of poisonous waste that fill the dead lake in Baotou. This is just one of the places that are brimming with what environmental studies scholar Myra Hird calls "the waste we want to forget."[36]

To date, the unique electronic, optical, and magnetic uses of rare earth elements cannot be matched by any other metals, but the ratio of usable minerals to waste toxins is extreme. Natural resource strategist David Abraham describes the mining in Jiangxi, China, of dysprosium and terbium, which are used in a variety of high-tech devices. He writes, "Only 0.2

percent of the mined clay contains the valuable rare earth elements. This means that 99.8 percent of earth removed in rare earth mining is discarded as waste, called 'tailings,' that are dumped back into the hills and streams," creating new pollutants like ammonium.[37] In order to refine one ton of these rare earth elements, "the Chinese Society of Rare Earths estimates that the process produces 75,000 liters of acidic water and one ton of radioactive residue."[38]

About three thousand miles south of Baotou are the small Indonesian islands of Bangka and Belitung, off the coast of Sumatra. Bangka and Belitung produce 90 percent of Indonesia's tin, used in semiconductors. Indonesia is the world's second-largest producer of the metal, behind China. Indonesia's national tin corporation, PT Timah, supplies companies such as Samsung directly, as well as solder makers Chernan and Shenmao, which in turn supply Sony, LG, and Foxconn— all suppliers for Apple, Tesla, and Amazon.[39]

On these small islands, gray-market miners who are not officially employed sit on makeshift pontoons, using bamboo poles to scrape the seabed before diving underwater to suck tin from the surface by drawing their breath through giant, vacuumlike tubes. The miners sell the tin they find to middlemen, who also collect ore from miners working in authorized mines, and they mix it together to sell to companies like Timah.[40] Completely unregulated, the process unfolds beyond any formal worker or environmental protections. As investigative journalist Kate Hodal reports, "Tin mining is a lucrative but destructive trade that has scarred the island's landscape, bulldozed its farms and forests, killed off its fish stocks and coral reefs, and dented tourism to its pretty palm-lined beaches. The damage is best seen from the air, as pockets of lush forest huddle amid huge swaths of barren orange earth. Where not dominated by mines, this is pockmarked with graves, many holding the bodies of miners who have died over the centuries digging for tin."[41] The mines are everywhere: in backyards, in the forest, by the side of the road, on the beaches. It is a landscape of ruin.

It is a common practice of life to focus on the world immediately before us, the one we see and smell and touch every day. It grounds us where we are, with our communities and our known corners and concerns. But to see the full supply chains of AI requires looking for patterns in a global sweep, a sensitivity to the ways in which the histories and specific harms are different from place to place and yet are deeply interconnected by the multiple forces of extraction.

We can see these patterns across space, but we can also find them across time. Transatlantic telegraph cables are the essential infrastructure that ferries data between the continents, an emblem of global communication and capital. They are also a material product of colonialism, with its patterns of extraction, conflict, and environmental destruction. At the end of the nineteenth century, a particular Southeast Asian tree called *Palaquium gutta* became the center of a cable boom. These trees, found mainly in Malaysia, produce a milky white natural latex called gutta-percha. After English scientist Michael Faraday published a study in the *Philosophical Magazine* in 1848 about the use of this material as an electrical insulator, gutta-percha rapidly became the darling of the engineering world. Engineers saw gutta-percha as the solution to the problem of insulating telegraphic cables to withstand harsh and varying conditions on the ocean floor. The twisted strands of copper wire needed four layers of the soft, organic tree sap to protect them from water incursion and carry their electrical currents.

As the global submarine telegraphy business grew, so did demand for *Palaquium gutta* tree trunks. The historian John Tully describes how local Malay, Chinese, and Dayak workers were paid little for the dangerous work of felling the trees and slowly collecting the latex.[42] The latex was processed and then sold through Singapore's trade markets into the British market, where it was transformed into, among other things, lengths upon lengths of submarine cable sheaths that wrapped around the globe. As media scholar Nicole Starosielski writes, "Military strategists saw cables as the most efficient and secure mode of communication with the colonies—and, by implication, of control over them."[43] The routes of submarine cables today still mark out the early colonial networks between the centers and the peripheries of empire.[44]

A mature *Palaquium gutta* could yield around eleven ounces of latex. But in 1857, the first transatlantic cable was around eighteen hundred miles long and weighed two thousand tons—requiring about 250 tons of gutta-percha. To produce just one ton of this material required around nine hundred thousand tree trunks. The jungles of Malaysia and Singapore were stripped; by the early 1880s, the *Palaquium gutta* had vanished. In a last-ditch effort to save their supply chain, the British passed a ban in 1883 to halt harvesting the latex, but the tree was all but extinct.[45]

The Victorian environmental disaster of gutta-percha, at the dawn of the global information society, shows how the relations between technology and its materials, environments, and labor practices are interwoven.[46] Just as Victorians precipitated ecological disaster for their early cables, so do contemporary mining and global supply chains further imperil the delicate ecological balance of our era.

There are dark ironies in the prehistories of planetary computation. Currently large-scale AI systems are driving forms of environmental, data, and human extraction, but from the Victorian era onward, algorithmic computation emerged out of desires to manage and control war, population, and cli-

Palaquium gutta

mate change. The historian Theodora Dryer describes how the
founding figure of mathematical statistics, English scientist
Karl Pearson, sought to resolve uncertainties of planning and
management by developing new data architectures including
standard deviations and techniques of correlation and regression. His methods were, in turn, deeply imbricated with race
science, as Pearson—along with his mentor, the statistician
and founder of eugenics Sir Francis Galton—believed that
statistics could be "the first step in an enquiry into the possible
effect of a selective process upon any character of a race."[47]

As Dryer writes, "By the end of the 1930s, these data architectures—regression techniques, standard deviation, and correlations—would become dominant tools used in interpreting
social and state information on the world stage. Tracking the
nodes and routes of global trade, the interwar 'mathematical-

statistics movement' became a vast enterprise."[48] This enterprise kept expanding after World War II, as new computational systems were used in domains such as weather forecasting during periods of drought to eke out more productivity from large-scale industrial farming.[49] From this perspective, algorithmic computing, computational statistics, and artificial intelligence were developed in the twentieth century to address social and environmental challenges but would later be used to intensify industrial extraction and exploitation and further deplete environmental resources.

The Myth of Clean Tech

Minerals are the backbone of AI, but its lifeblood is still electrical energy. Advanced computation is rarely considered in terms of carbon footprints, fossil fuels, and pollution; metaphors like "the cloud" imply something floating and delicate within a natural, green industry.[50] Servers are hidden in nondescript data centers, and their polluting qualities are far less visible than the billowing smokestacks of coal-fired power stations. The tech sector heavily publicizes its environmental policies, sustainability initiatives, and plans to address climate-related problems using AI as a problem-solving tool. It is all part of a highly produced public image of a sustainable tech industry with no carbon emissions. In reality, it takes a gargantuan amount of energy to run the computational infrastructures of Amazon Web Services or Microsoft's Azure, and the carbon footprint of the AI systems that run on those platforms is growing.[51]

As Tung-Hui Hu writes in *A Prehistory of the Cloud*, "The cloud is a resource-intensive, extractive technology that converts water and electricity into computational power, leaving a sizable amount of environmental damage that it then displaces from sight."[52] Addressing this energy-intensive infrastructure has become a major concern. Certainly, the industry has made significant efforts to make data centers more energy efficient and to increase their use of renewable energy. But already, the carbon footprint of the world's computational infrastructure has matched that of the aviation industry at its height, and it is increasing at a faster rate.[53] Estimates vary, with researchers like Lotfi Belkhir and Ahmed Elmeligi estimating that the tech sector will contribute 14 percent of global greenhouse emissions by 2040, while a team in Sweden predicts that the electricity demands of data centers alone will increase about fifteenfold by 2030.[54]

By looking closely at the computational capacity needed to build AI models, we can see how the desire for exponential increases in speed and accuracy is coming at a high cost to the planet. The processing demands of training AI models, and thus their energy consumption, is still an emerging area of investigation. One of the early papers in this field came from AI researcher Emma Strubell and her team at the University of Massachusetts Amherst in 2019. With a focus on trying to understand the carbon footprint of natural language processing (NLP) models, they began to sketch out potential estimates by running AI models over hundreds of thousands of computational hours.[55] The initial numbers were striking. Strubell's team found that running only a single NLP model produced more than 660,000 pounds of carbon dioxide emissions, the equivalent of five gas-powered cars over their total lifetime (including their manufacturing) or 125 round-trip flights from New York to Beijing.[56]

Worse, the researchers noted that this modeling is, at minimum, a baseline optimistic estimate. It does not reflect the true commercial scale at which companies like Apple and Amazon operate, scraping internet-wide datasets and feeding their own NLP models to make AI systems like Siri and Alexa sound more human. But the exact amount of energy consumption produced by the tech sector's AI models is unknown; that information is kept as highly guarded corporate secrets. Here, too, the data economy is premised on maintaining environmental ignorance.

In the AI field, it is standard practice to maximize computational cycles to improve performance, in accordance with a belief that bigger is better. As Rich Sutton of DeepMind describes it: "Methods that leverage computation are ultimately the most effective, and by a large margin."[57] The computational technique of brute-force testing in AI training runs, or systematically gathering more data and using more computational cycles until a better result is achieved, has driven a steep increase in energy consumption. OpenAI estimated that since 2012, the amount of compute used to train a single AI model has increased by a factor of ten every year. That's due to developers "repeatedly finding ways to use more chips in parallel, and being willing to pay the economic cost of doing so."[58] Thinking only in terms of economic cost narrows the view on the wider local and environmental price of burning computation cycles as a way to create incremental efficiencies. The tendency toward "compute maximalism" has profound ecological impacts.

Data centers are among the world's largest consumers of electricity.[59] Powering this multilevel machine requires grid electricity in the form of coal, gas, nuclear, or renewable energy. Some corporations are responding to growing alarm about the energy consumption of large-scale computation, with Apple and Google claiming to be carbon neutral (which means they offset their carbon emissions by purchasing credits) and Microsoft promising to become carbon negative by 2030. But workers within the companies have pushed for reductions in emissions across the board, rather than what they see as buying indulgences out of environmental guilt.[60] Moreover, Microsoft, Google, and Amazon all license their AI platforms, engineering workforces, and infrastructures to fossil fuel companies to help them locate and extract fuel from the ground, which further drives the industry most responsible for anthropogenic climate change.

Beyond the United States, more clouds of carbon dioxide are rising. China's data center industry draws 73 percent of its power from coal, emitting about 99 million tons of CO_2 in 2018.[61] And electricity consumption from China's data center infrastructure is expected to increase by two-thirds by 2023.[62] Greenpeace has raised the alarm about the colossal energy demands of China's biggest technology companies, arguing that "China's leading tech companies, including Alibaba, Tencent, and GDS, must dramatically scale up clean energy procurement and disclose energy use data."[63] But the lasting impacts of coal-fired power are everywhere, exceeding any national boundaries. The planetary nature of resource extraction and its consequences goes well beyond what the nation-state was designed to address.

Water tells another story of computation's true cost. The history of water use in the United States is full of battles and secret deals, and as with computation, the deals made over water are kept close. One of the biggest U.S. data centers belongs to the National Security Agency (NSA) in Bluffdale, Utah. Open since late 2013, the Intelligence Community Comprehensive National Cybersecurity Initiative Data Center is impossible to visit directly. But by driving up through the adjacent suburbs, I found a cul-de-sac on a hill thick with sagebrush, and from there I was afforded a closer view of the sprawling 1.2-million-square-foot facility. The site has a kind of symbolic power of the next era of government data capture, having been featured in films like *Citizenfour* and pictured in thousands of news stories about the NSA. In person, though, it looks non-descript and prosaic, a giant storage container combined with a government office block.

The struggle over water began even before the data center was officially open, given its location in drought-parched Utah.[64] Local journalists wanted to confirm whether the estimated consumption of 1.7 million gallons of water per day was accurate, but the NSA initially refused to share usage data, redacted all details from public records, and claimed that its water use was a matter of national security. Antisurveillance activists created handbooks encouraging the end of material support of water and energy to surveillance, and they strategized that legal controls over water usage could help shut down the facility.[65] But the city of Bluffdale had already made a multiyear deal with the NSA, in which the city would sell water at rates well below the average in return for the promise of economic growth the facility might bring to the region.[66] The geopolitics of water are now deeply combined with the mechanisms and politics of data centers, computation, and power—in every sense. From the dry hillside that overlooks the NSA's data repository, all the contestation and obfuscation about water makes sense: this is a landscape with a limit, and water that is used to cool servers is being taken away from communities and habitats that rely on it to live.

Just as the dirty work of the mining sector was far removed from the companies and city dwellers who profited most, so the majority of data centers are far removed from major population hubs, whether in the desert or in semi-industrial exurbs. This contributes to our sense of the cloud being out of sight and abstracted away, when in fact it is material, affecting the environment and climate in ways that are far from being fully recognized and accounted for. The cloud is of the earth, and to keep it growing requires expanding resources and layers of logistics and transport that are in constant motion.

The Logistical Layer

So far, we have considered the material stuff of AI, from rare earth elements to energy. By grounding our analysis in the specific materialities of AI—the things, places, and people—we can better see how the parts are operating within broader systems of power. Take, for example, the global logistical machines that move minerals, fuel, hardware, workers, and consumer AI devices around the planet.[67] The dizzying spectacle of logistics and production displayed by companies like Amazon would not be possible without the development and widespread acceptance of a standardized metal object: the cargo container. Like submarine cables, cargo containers bind the industries of global communication, transport, and capital, a material exercise of what mathematicians call "optimal transport"—in this case, as an optimization of space and resources across the trade routes of the world.

Standardized cargo containers (themselves built from the basic earth elements of carbon and iron forged as steel) enabled the explosion of the modern shipping industry, which in turn made it possible to envision and model the planet as a single massive factory. The cargo container is the single unit of value—like a piece of Lego—that can travel thousands of miles before meeting its final destination as a modular part of a greater system of delivery. In 2017, the capacity of container ships in seaborne trade reached nearly 250 million deadweight tons of cargo, dominated by giant shipping companies including Maersk of Denmark, the Mediterranean Shipping Company of Switzerland, and France's CMA CGM Group, each

owning hundreds of container vessels.[68] For these commercial
ventures, cargo shipping is a relatively cheap way to navigate
the vascular system of the global factory, yet it disguises far
larger external costs. Just as they tend to neglect the physical
realities and costs of AI infrastructure, popular culture and
media rarely cover the shipping industry. The author Rose
George calls this condition "sea blindness."[69]

In recent years, shipping vessels produced 3.1 percent of
yearly global carbon dioxide emissions, more than the total
produced by Germany.[70] In order to minimize their internal
costs, most container shipping companies use low-grade fuel
in enormous quantities, which leads to increased amounts of
airborne sulfur and other toxic substances. One container ship
is estimated to emit as much pollution as fifty million cars, and
sixty thousand deaths every year are attributed indirectly to
cargo-ship-industry pollution.[71]

Even industry-friendly sources like the World Shipping
Council admit that thousands of containers are lost each year,
sinking to the ocean floor or drifting loose.[72] Some carry toxic
substances that leak into the oceans; others release thousands
of yellow rubber ducks that wash ashore around the world over
decades.[73] Typically, workers spend almost six months at sea,
often with long working shifts and without access to external
communications.

Here, too, the most severe costs of global logistics are
borne by the Earth's atmosphere, the oceanic ecosystem and
low-paid workers. The corporate imaginaries of AI fail to depict the lasting costs and long histories of the materials needed
to build computational infrastructures or the energy required
to power them. The rapid growth of cloud-based computation, portrayed as environmentally friendly, has paradoxically
driven an expansion of the frontiers of resource extraction. It
is only by factoring in these hidden costs, these wider collec-

tions of actors and systems, that we can understand what the shift toward increasing automation will mean. This requires working against the grain of how the technological imaginary usually works, which is completely untethered from earthly matters. Like running an image search of "AI," which returns dozens of pictures of glowing brains and blue-tinted binary code floating in space, there is a powerful resistance to engaging with the materialities of these technologies. Instead, we begin with the earth, with extraction, and with the histories of industrial power and then consider how these patterns are repeated in systems of labor and data.

AI as Megamachine

In the late 1960s, the historian and philosopher of technology Lewis Mumford developed the concept of the *megamachine* to illustrate how all systems, no matter how immense, consist of the work of many individual human actors.[74] For Mumford, the Manhattan Project was the defining modern megamachine whose intricacies were kept not only from the public but even from the thousands of people who worked on it at discrete, secured sites across the United States. A total of 130,000 workers operated in complete secrecy under the direction of the military, developing a weapon that would kill (by conservative estimates) 237,000 people when it hit Hiroshima and Nagasaki in 1945. The atomic bomb depended on a complex, secret chain of supply, logistics, and human labor.

Artificial intelligence is another kind of megamachine, a set of technological approaches that depend on industrial infrastructures, supply chains, and human labor that stretch around the globe but are kept opaque. We have seen how AI is much more than databases and algorithms, machine learning models and linear algebra. It is metamorphic: relying on manufacturing, transportation, and physical work; data centers and the undersea cables that trace lines between the continents; personal devices and their raw components; transmission signals passing through the air; datasets produced by scraping the internet; and continual computational cycles. These all come at a cost.

We have looked at the relations between cities and mines, companies and supply chains, and the topographies of extraction that connect them. The fundamentally intertwined nature of production, manufacturing, and logistics reminds us that the mines that drive AI are everywhere: not only sited in discrete locations but diffuse and scattered across the geography of the earth, in what Mazen Labban has called the "planetary mine."[75] This is not to deny the many specific locations where technologically driven mining is taking place. Rather, Labban observes that the planetary mine expands and reconstitutes extraction into novel arrangements, extending the practices of mines into new spaces and interactions around the world.

Finding fresh methods for understanding the deep material and human roots of AI systems is vital at this moment in history, when the impacts of anthropogenic climate change are already well under way. But that's easier said than done. In part, that's because many industries that make up the AI system chain conceal the ongoing costs of what they do. Furthermore, the scale required to build artificial intelligence systems is too complex, too obscured by intellectual property law, and too mired in logistical and technical complexity for us to see into it all. But the aim here is not to try and make these complex assemblages transparent: rather than trying to see *inside* them, we will be connecting *across* multiple systems to understand how they work in relation to each other.[76] Thus, our path

The ruins at Blair. Photograph by Kate Crawford

will follow the stories about the environmental and labor costs of AI and place them in context with the practices of extraction and classification braided throughout everyday life. It is by thinking about these issues together that we can work toward greater justice.

I make one more trip to Silver Peak. Before I reach the town, I pull the van over to the side of the road to read a weather-beaten sign. It's Nevada Historical Marker 174, dedicated to the creation and destruction of a small town called Blair. In 1906, the Pittsburgh Silver Peak Gold Mining Company bought up the mines in the area. Anticipating a boom, land speculators purchased all of the available plots near Silver Peak along with its water rights, driving prices to record artificial highs. So the mining company surveyed a couple of miles north and declared it the site for a new town: Blair. They built a hundred-stamp cyanide mill for leach mining, the biggest in the state, and laid the Silver Peak railroad that ran from Blair

Junction to the Tonopah and Goldfield main line. Briefly, the town thrived. Many hundreds of people came from all over for the jobs, despite the harsh working conditions. But with so much mining activity, the cyanide began to poison the ground, and the gold and silver seams began to falter and dry up. By 1918, Blair was all but deserted. It was all over within twelve years. The ruins are marked on a local map—just a forty-five-minute walk away.

It's a blazing hot day in the desert. The only sounds are the metallic reverberations of cicadas and the rumble of an occasional passenger jet. I decide to start up the hill. By the time I reach the collection of stone buildings at the top of the long dirt road, I'm exhausted from the heat. I take shelter inside the collapsed remains of what was once a gold miner's house. Not much is left: some broken crockery, shards of glass bottles, a few rusted tins. Back in Blair's lively years, multiple saloons thrived nearby and a two-story hotel welcomed visitors. Now it's a cluster of broken foundations.

Through the space where a window used to be, the view stretches all the way down the valley. I'm struck by the realization that Silver Peak will also be a ghost town soon. The current draw on the lithium mine is aggressive in response to the high demand, and no one knows how long it will last. The most optimistic estimate is forty years, but the end may come much sooner. Then the lithium pools under the Clayton Valley will be exsanguinated—extracted for batteries that are destined for landfill. And Silver Peak will return to its previous life as an empty and quiet place, on the edge of an ancient salt lake, now drained.

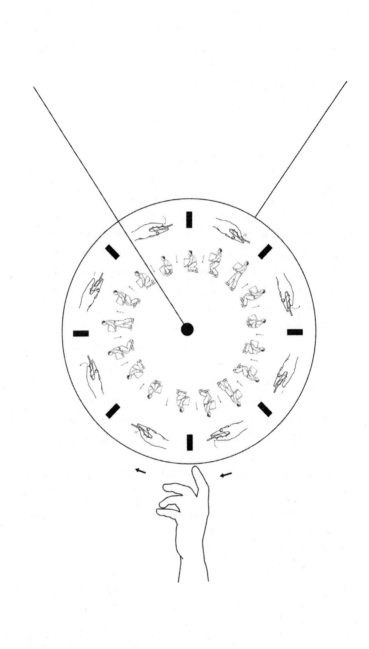

2

Labor

When I enter Amazon's vast fulfillment center in Robbinsville, New Jersey, the first thing I see is a large sign that reads "Time Clock." It juts out from one of the bright yellow concrete pylons spanning across the vast factory space of 1.2 million square feet. This is a major distribution warehouse for smaller objects—a central distribution node for the Northeastern United States. It presents a dizzying spectacle of contemporary logistics and standardization, designed to accelerate the delivery of packages. Dozens of time-clock signs appear at regular intervals along the entryway. Every second of work is being monitored and tallied. Workers—known as "associates"—must scan themselves in as soon as they arrive. The sparse, fluorescent-lit break rooms also feature time clocks—with more signs to underscore that all scans in and out of the rooms are tracked. Just as packages are scanned in the warehouse, so too are workers monitored for the greatest possible efficiency: they can only be off-task for fifteen minutes per shift, with an unpaid thirty-minute meal break. Shifts are ten hours long.

This is one of the newer fulfillment centers that feature

robots to move the heavy shelving units laden with products in trays. The bright orange Kiva robots glide smoothly across the concrete floors like vivid water bugs, following a programmed logic that causes them to spin in lazy circles and then lock onto a path toward the next worker awaiting the trays. Then they move forward, carrying on their backs a tower of purchases that can weigh up to three thousand pounds. This shuffling army of ground-hugging robots presents a kind of effortless efficiency: they carry, they rotate, they advance, they repeat. They make a low, whirring hum, but it is almost entirely drowned out by the deafening sound of fast-moving conveyor belts that act as the factory's arteries. There are fourteen miles of conveyor belts moving without pause in this space. The result is a constant roar.

While the robots perform their coordinated algorithmic ballet behind bare chain-link fences, the workers in the factory are far less serene. The anxiety of making the "picking rate"—the number of items they must select and pack within the allocated time—is clearly taking a toll. Many of the workers I encounter on my visits are wearing some kind of support bandage. I see knee braces, elbow bandages, wrist guards. When I observe that many people seem to have injuries, the Amazon worker guiding me through the factory points to the vending machines spaced at regular intervals that are "stocked with over-the-counter painkillers for anyone who needs them."

Robotics has become a key part of Amazon's logistical armory, and while the machinery seems well tended, the corresponding human bodies seem like an afterthought. They are there to complete the specific, fiddly tasks that robots cannot: picking up and visually confirming all of the oddly shaped objects that people want delivered to their homes, from phone cases to dishwashing detergent, within the shortest amount of time. Humans are the necessary connective tissue to get

Workers and time clocks at the Amazon fulfillment center in
Robbinsville Township, New Jersey. AP Photo/Julio Cortez

ordered items into containers and trucks and delivered to consumers. But they aren't the most valuable or trusted component of Amazon's machine. At the end of the day, all associates
must exit through a row of metal detectors. This is an effective
antitheft measure, I am told.

Within the layers of the internet, one of the most common units of measurement is the network packet—a basic
unit of data to be sent from one destination and delivered to
another. At Amazon, the basic unit of measurement is the
brown cardboard box, that familiar domestic cargo vessel em-
blazoned with a curved arrow simulating a human smile. Network packets each have a timestamp known as a *time to live*.
Data has to reach its destination before the time to live expires.
At Amazon, the cardboard box also has a *time to live* driven

by the customer's shipping demands. If the box is late, this affects Amazon's brand and ultimately its profits. So enormous attention has been devoted to the machine learning algorithm that is tuned to the data regarding the best size, weight, and strength of corrugated boxes and paper mailers. Apparently without irony, the algorithm is called "the matrix."[1] Whenever a person reports a broken item, it becomes a data point about what sort of box should be used in the future. The next time that product is mailed, it will automatically be assigned a new type of box by the matrix, without human input. This prevents breakages, which saves time, which increases profits. Workers, however, are forced continually to adapt, which makes it harder to put their knowledge into action or habituate to the job.

The control over time is a consistent theme in the Amazon logistical empire, and the bodies of workers are run according to the cadences of computational logics. Amazon is America's second-largest private employer, and many companies strive to emulate its approach. Many large corporations are heavily investing in automated systems in the attempt to extract ever-larger volumes of labor from fewer workers. Logics of efficiency, surveillance, and automation are all converging in the current turn to computational approaches to managing labor. The hybrid human-robotic distribution warehouses of Amazon are a key site to understand the trade-offs being made in this commitment to automated efficiency. From there, we can begin to consider the question of how labor, capital, and time are entwined in AI systems.

Rather than debating whether humans will be replaced by robots, in this chapter I focus on how the experience of work is shifting in relation to increased surveillance, algorithmic assessment, and the modulation of time. Put another way, instead of asking whether robots will replace humans, I'm interested in how humans are increasingly treated like robots and what this means for the role of labor. Many forms of work are shrouded in the term "artificial intelligence," hiding the fact that people are often performing rote tasks to shore up the impression that machines can do the work. But large-scale computation is deeply rooted in and running on the exploitation of human bodies.

If we want to understand the future of work in the context of artificial intelligence, we need to begin by understanding the past and present experience of workers. Approaches to maximizing the extraction of value from workers vary from reworkings of the classical techniques used in Henry Ford's factories to a range of machine learning–assisted tools designed to increase the granularity of tracking, nudging, and assessment. This chapter maps geographies of labor past and present, from Samuel Bentham's inspection houses to Charles Babbage's theories of time management and to Frederick Winslow Taylor's micromanagement of human bodies. Along the way, we will see how AI is built on the very human efforts of (among other things) crowdwork, the privatization of time, and the seemingly never-ending reaching, lifting, and toiling of putting boxes into order. From the lineage of the mechanized factory, a model emerges that values increased conformity, standardization, and interoperability—for products, processes, and humans alike.

Prehistories of Workplace AI

Workplace automation, though often told as a story of the future, is already a long-established experience of contemporary work. The manufacturing assembly line, with its emphasis on consistent and standardized units of production, has analogues in the service industries, from retail to restaurants. Secretarial labor has been increasingly automated since the 1980s and now is emulated by highly feminized AI assistants such as Siri, Cortana, and Alexa.[2] So-called knowledge workers, those white-collar employees assumed to be less threatened by the forces driving automation, find themselves increasingly subjected to workplace surveillance, process automation, and collapse between the distinction of work and leisure time (although women have rarely experienced such clear distinctions, as feminist theorists of work like Silvia Federici and Melissa Gregg have shown).[3] Work of all stripes has had to significantly adapt itself in order to be interpretable and understood by software-based systems.[4]

The common refrain for the expansion of AI systems and process automation is that we are living in a time of beneficial human-AI collaboration. But this collaboration is not fairly negotiated. The terms are based on a significant power asymmetry—is there ever a choice *not* to collaborate with algorithmic systems? When a company introduces a new AI platform, workers are rarely allowed to opt out. This is less of a collaboration than a forced engagement, where workers are expected to re-skill, keep up, and unquestioningly accept each new technical development.

Rather than representing a radical shift from established forms of work, the encroachment of AI into the workplace should properly be understood as a return to older practices of industrial labor exploitation that were well established in the 1890s and the early twentieth century. That was a time when factory labor was already seen in relation to machines and work tasks were increasingly subdivided into smaller actions requiring minimal skill but maximum exertion. Indeed, the current expansion of labor automation continues the broader historical dynamics inherent in industrial capitalism. Since the appearance of the earliest factories, workers have encountered ever more powerful tools, machines, and electronic systems that play a role in changing how labor is managed while transferring more value to their employers. We are witnessing new refrains on an old theme. The crucial difference is that employers now observe, assess, and modulate intimate parts of the work cycle and bodily data—down to the last micromovement—that were previously off-limits to them.

There are many prehistories of workplace AI; one is the Industrial Revolution's widespread automation of common productive activities. In his *Wealth of Nations,* the eighteenth-century political economist Adam Smith first pointed to the division and subdivision of manufacturing tasks as the basis of both improved productivity and increasing mechanization.[5] He observed that by identifying and analyzing the various steps involved in manufacturing any given item, it was possible to divide them into ever-smaller steps, so that a product once made entirely by expert craftspeople could now be built by a team of lower-skill workers equipped with tools purpose-built for a particular task. Thus, a factory's output could be scaled up significantly without an equivalent increase in labor cost.

Developments in mechanization were important, but it was only when combined with a growing abundance of energy derived from fossil fuels that they could drive a massive increase in the productive capacities of industrial societies. This increase in production occurred in tandem with a major transformation of the role of labor vis-à-vis machinery in the workplace. Initially conceived as labor-saving devices, factory machines were meant to assist workers with their daily activities but quickly became the center of productive activity, shaping the speed and character of work. Steam engines powered by coal and oil could drive continuous mechanical actions that influenced the pace of work in the factory. Work ceased to be primarily a product of human labor and took on an increasingly machinelike character, with workers adapting to the needs of the machine and its particular rhythms and cadences. Building on Smith, Karl Marx noted as early as 1848 that automation abstracts labor from the production of finished objects and turns a worker into "an appendage of the machine."[6]

The integration of workers' bodies with machines was sufficiently thorough that early industrialists could view their employees as a raw material to be managed and controlled like any other resource. Factory owners, using both their local political clout and paid muscle, sought to direct and restrict how their workers moved around within factory towns, sometimes even preventing workers from emigrating to less mechanized regions of the world.[7]

This also meant increasing control over time. The historian E. P. Thompson's formative essay explores how the Industrial Revolution demanded greater synchronization of work and stricter time disciplines.[8] The transition to industrial capitalism came with new divisions of labor, oversight, clocks, fines, and time sheets—technologies that also influenced the way people experienced time. Culture was also a powerful tool. During the eighteenth and nineteenth centuries, the propaganda about hard work came in the forms of pamphlets and essays on the importance of discipline and sermons on the virtues of early rising and working diligently for as long as possible.[9] The use of time came to be seen in both moral and economic terms: understood as a currency, time could be well spent or squandered away. But as more rigid time disciplines were imposed in workshops and factories, the more workers began to push back—campaigning over time itself. By the 1800s, labor movements were strongly advocating for reducing the working day, which could run as long as sixteen hours. Time itself became a key site for struggle.

Maintaining an efficient and disciplined workforce in the early factory necessitated new systems of surveillance and

control. One such invention from the early years of industrial manufacturing was the inspection house, a circular arrangement that placed all of a factory's workers within sight
of their supervisors, who worked from an office placed on a
raised platform at the center of the building. Developed in the
1780s in Russia by the English naval engineer Samuel Bentham
while under the employ of Prince Potemkin, this arrangement
allowed expert supervisors to keep an eye on their untrained
subordinates—mostly Russian peasants loaned to Bentham by
Potemkin—for signs of poor working habits. It also allowed
Bentham himself to keep an eye on the supervisors for signs of
ill-discipline. The supervisors, mostly master shipbuilders recruited from England, caused Bentham great annoyance due to
their tendency to drink and get into petty disagreements with
one another. "Morning after morning I am taken up chiefly
with disputes amongst my Officers," Bentham complained.[10]
As his frustrations grew, he embarked on a redesign that would
maximize his ability to keep a watchful eye on them, and on
the system as a whole. With a visit from his elder brother, the
utilitarian philosopher Jeremy Bentham, Samuel's inspection
house became the inspiration for the famous panopticon, a
design for a model prison featuring a central watchtower from
which guards could supervise the prisoners in their cells.[11]

Since Michel Foucault's *Discipline and Punish*, it has become commonplace to consider the prison as the origin point
of today's surveillance society, with the elder Bentham as its
ideological progenitor. In fact, the panoptic prison owes its
origins to the work of the younger Bentham in the context of
the early manufacturing facility.[12] The panopticon began as a
workplace mechanism well before it was conceptualized for
prisons.

While Samuel Bentham's work on the inspection house
has largely faded from our collective memory, the story behind

it remains part of our shared lexicon. The inspection house was part of a strategy coordinated by Bentham's employer, Prince Potemkin, who wished to gain favor in Catherine the Great's court by demonstrating the potential for modernizing rural Russia and transforming the peasantry into a modern manufacturing workforce. The inspection house was built to serve as a spectacle for visiting dignitaries and financiers, much like the so-called Potemkin villages, which were little more than decorated facades designed to distract observers from the impoverished rural village landscapes discreetly obscured from view.

And this is just one genealogy. Many other histories of labor shaped these practices of observation and control. The plantation colonies of the Americas used forced labor to maintain cash crops like sugar, and slave owners depended on systems of constant surveillance. As Nicholas Mirzoeff describes in *The Right to Look,* a central role in the plantation economy was the overseer, who watched over the flow of production on the colonial slave plantation, and their oversight meant ordering the work of the slaves within a system of extreme violence.[13] As one planter described in 1814, the role of the overseer was "to never leave the slave for an instant in inaction; he keeps the fabrication of sugar under surveillance, never leaving the sugar-mill for an instant."[14] This regime of oversight also relied on bribing some slaves with food and clothing to enlist them as an expanded surveillance network and to maintain discipline and speed of work when the overseer was occupied.[15]

Now the role of oversight in the modern workplace is primarily deputized to surveillance technologies. The managerial class employs a wide range of technologies to surveil employees, including tracking their movements with apps, analyzing their social media feeds, comparing the patterns of replying to emails and booking meetings, and nudging them with suggestions to make them work faster and more efficiently. Employee data is used to make predictions about who is most likely to succeed (according to narrow, quantifiable parameters), who might be diverging from company goals, and who might be organizing other workers. Some use the techniques of machine learning, and others are more simplistic algorithmic systems. As workplace AI becomes more prevalent, many of the more basic monitoring and tracking systems are being expanded with new predictive capacities to become increasingly invasive mechanisms of worker management, asset control, and value extraction.

Potemkin AI and the Mechanical Turks

One of the less recognized facts of artificial intelligence is how many underpaid workers are required to help build, maintain, and test AI systems. This unseen labor takes many forms—supply-chain work, on-demand crowdwork, and traditional service-industry jobs. Exploitative forms of work exist at all stages of the AI pipeline, from the mining sector, where resources are extracted and transported to create the core infrastructure of AI systems, to the software side, where distributed workforces are paid pennies per microtask. Mary Gray and Sid Suri refer to such hidden labor as "ghost work."[16] Lilly Irani calls it "human-fueled automation."[17] These scholars have drawn attention to the experiences of crowdworkers or microworkers who perform the repetitive digital tasks that underlie AI systems, such as labeling thousands of hours of training data and reviewing suspicious or harmful content. Workers do the repetitive tasks that backstop claims of AI magic—but they rarely receive credit for making the systems function.[18]

Although this labor is essential to sustaining AI systems, it is usually very poorly compensated. A study from the United Nations International Labour Organization surveyed

3,500 crowdworkers from seventy-five countries who routinely offered their labor on popular task platforms like Amazon Mechanical Turk, Figure Eight, Microworkers, and Clickworker. The report found that a substantial number of people earned below their local minimum wage even though the majority of respondents were highly educated, often with specializations in science and technology.[19] Likewise, those who do content moderation work—assessing violent videos, hate speech, and forms of online cruelty for deletion—are also paid poorly. As media scholars such as Sarah Roberts and Tarleton Gillespie have shown, this kind of work can leave lasting forms of psychological trauma.[20]

But without this kind of work, AI systems won't function. The technical AI research community relies on cheap, crowd-sourced labor for many tasks that can't be done by machines. Between 2008 and 2016, the term *"crowdsourcing"* went from appearing in fewer than a thousand scientific articles to more than twenty thousand—which makes sense, given that Mechanical Turk launched in 2005. But during the same time frame, there was far too little debate about what ethical questions might be posed by relying on a workforce that is commonly paid far below the minimum wage.[21]

Of course, there are strong incentives to ignore the dependency on underpaid labor from around the world. All the work they do—from tagging images for computer-vision systems to testing whether an algorithm is producing the right results—refines AI systems much more quickly and cheaply, particularly when compared to paying students to do these tasks (as was the earlier tradition). So the issue has generally been ignored, and as one crowdwork research team observed, clients using these platforms "expect cheap, 'frictionless' completion of work without oversight, as if the platform were not an interface to human workers but a vast computer without living expenses."[22] In other words, clients treat human employees as little more than machines, because to recognize their work and compensate it fairly would make AI more expensive and less "efficient."

Sometimes workers are directly asked to pretend to be an AI system. The digital personal assistant start-up x.ai claimed that its AI agent, called Amy, could "magically schedule meetings" and handle many mundane daily tasks. But a detailed Bloomberg investigation by journalist Ellen Huet revealed that it wasn't artificial intelligence at all. "Amy" was carefully being checked and rewritten by a team of contract workers pulling long shifts. Similarly, Facebook's personal assistant, M, was relying on regular human intervention by a group of workers paid to review and edit every message.[23]

Faking AI is an exhausting job. The workers at x.ai were sometimes putting in fourteen-hour shifts of annotating emails in order to sustain the illusion that the service was automated and functioning 24/7. They couldn't leave at the end of the night until the queues of emails were finished. "I left feeling totally numb and absent of any sort of emotion," one employee told Huet.[24]

We could think of this as a kind of Potemkin AI—little more than facades, designed to demonstrate to investors and a credulous media what an automated system would look like while actually relying on human labor in the background.[25] In a charitable reading, these facades are an illustration of what the system might be capable of when fully realized, or a "minimum viable product" designed to demonstrate a concept. In a less charitable reading, Potemkin AI systems are a form of deception perpetrated by technology vendors eager to stake a claim in the lucrative tech space. But until there is another way to create large-scale AI that doesn't use extensive behind-the-curtain work by humans, this is a core logic of how AI works.

The writer Astra Taylor has described the kind of over-selling of high-tech systems that aren't actually automated as "fauxtomation."[26] Automated systems appear to do work previously performed by humans, but in fact the system merely coordinates human work in the background. Taylor cites the examples of self-service kiosks in fast-food restaurants and self-checkout systems in supermarkets as places where an employee's labor appears to have been replaced by an automated system but where in fact the data-entry labor has simply been relocated from a paid employee to the customer. Meanwhile, many online systems that provide seemingly automated decisions, such as removing duplicated entries or deleting offensive content, are actually powered by humans working from home on endless queues of mundane tasks.[27] Much like Potemkin's decorated villages and model workshops, many valuable automated systems feature a combination of underpaid digital pieceworkers and consumers taking on unpaid tasks to make systems function. Meanwhile, companies seek to convince investors and the general public that intelligent machines are doing the work.

What is at stake in this artifice? The true labor costs of AI are being consistently downplayed and glossed over, but the forces driving this performance run deeper than merely marketing trickery. It is part of a tradition of exploitation and deskilling, where people must do more tedious and repetitive work to back-fill for automated systems, for a result that may be less effective or reliable than what it replaced. But this approach can *scale*—producing cost reductions and profit increases while obscuring how much it depends on remote workers being paid subsistence wages and off-loading additional tasks of maintenance or error-checking to consumers.

Fauxtomation does not directly replace human labor; rather, it relocates and disperses it in space and time. In so doing it increases the disconnection between labor and value and thereby performs an ideological function. Workers, having been alienated from the results of their work as well as disconnected from other workers doing the same job, are liable to be more easily exploited by their employers. This is evident from the extremely low rates of compensation crowdworkers receive around the world.[28] They and other kinds of fauxtomation laborers face the very real fact that their labor is interchangeable with any of the thousands of other workers who compete with them for work on platforms. At any point they could be replaced by another crowdworker, or possibly by a more automated system.

In 1770, Hungarian inventor Wolfgang von Kempelen constructed an elaborate mechanical chess player. He built a cabinet of wood and clockwork, behind which was seated a life-size mechanical man who could play chess against human opponents and win. This extraordinary contraption was first shown in the court of Empress Maria Theresa of Austria, then to visiting dignitaries and government ministers, all of whom were utterly convinced that this was an intelligent automaton. The lifelike machine was dressed in a turban, wide-legged pants, and a fur-trimmed robe to give the impression of an "oriental sorcerer."[29] This racialized appearance signaled exotic otherness, at a time when the elites of Vienna would drink Turkish coffee and dress their servants in Turkish costumes.[30] It came to be known as the Mechanical Turk. But the chess-playing automaton was an elaborate illusion: it had a human chess master hiding inside an internal chamber, operating the machine from within and completely out of sight.

Some 250 years later, the hoax lives on. Amazon chose to name its micropayment-based crowdsourcing platform "Amazon Mechanical Turk," despite the association with racism and trickery. On Amazon's platform, real workers remain out of sight in service of an illusion that AI systems are autonomous and magically intelligent.[31] Amazon's initial motivation to build Mechanical Turk emerged from the failures of its own artificial intelligence systems that could not adequately detect duplicate product pages on its retail site. After a series of futile and expensive attempts to solve the problem, the project engineers enlisted humans to fill the gaps in its streamlined systems.[32] Now Mechanical Turk connects businesses with an unseen and anonymous mass of workers who bid against one another for the opportunity to work on a series of microtasks. Mechanical Turk is a massively distributed workshop where humans emulate and improve on AI systems by checking and correcting algorithmic processes. This is what Amazon chief executive Jeff Bezos brazenly calls "*artificial* artificial intelligence."[33]

These examples of Potemkin AI are all around. Some are directly visible: when we see one of the current crop of self-driving cars on the streets, we also see a human operator in the driver's seat, ready to take control of the vehicle at the first sign of trouble. Others are less visible, as when we interact with a web-based chat interface. We engage only with the facades that obscure their inner workings, designed to hide the various combinations of machine and human labor in each interaction. We aren't informed whether we are receiving a response from the system itself or from a human operator paid to respond on its behalf.

If there is growing uncertainty about whether we are engaging with an AI system or not, the feeling is mutual. In a paradox that many of us have experienced, and ostensibly in order to prove true human identity when reading a website, we are required to convince Google's reCAPTCHA of our humanity. So we dutifully select multiple boxes containing street numbers, or cars, or houses. We are training Google's image recognition algorithms for free. Again, the myth of AI as affordable and efficient depends on layers of exploitation, including the extraction of mass unpaid labor to fine-tune the AI systems of the richest companies on earth.

Contemporary forms of artificial intelligence are neither artificial nor intelligent. We can — and should — speak instead of the hard physical labor of mine workers, the repetitive factory labor on the assembly line, the cybernetic labor in the cognitive sweatshops of outsourced programmers, the poorly paid crowdsourced labor of Mechanical Turk workers, and the unpaid immaterial work of everyday users. These are the places where we can see how planetary computation depends on the exploitation of human labor, all along all the supply chain of extraction.

Visions of Disassembly and Workplace Automation: Babbage, Ford, and Taylor

Charles Babbage is well known as the inventor of the first mechanical computer. In the 1820s, he developed the idea for the *Difference Engine,* a mechanical calculating machine designed to generate mathematical and astronomical tables in a fraction of the time required to calculate them by hand. By the 1830s, he had a viable conceptual design for the *Analytical Engine,* a programmable general-purpose mechanical computer, complete with a system of punch cards for providing it with instructions.[34]

Babbage also had a strong interest in liberal social theory and wrote extensively on the nature of labor — the combination of his interests in computation and worker automation. Following Adam Smith, he noted the division of labor as a means of streamlining factory work and generating efficiencies. He went further, however, arguing that the industrial corporation could be understood as an analogue to a computational system. Just like a computer, it included multiple specialized units performing particular tasks, all coordinated to produce a given body of work, but with the labor content of the finished product rendered largely invisible by the process as a whole.

In Babbage's more speculative writing, he imagined perfect flows of work through the system that could be visualized as data tables and monitored by pedometers and repeating clocks.[35] Through a combination of computation, surveillance, and labor discipline, he argued, it would be possible to enforce ever-higher degrees of efficiency and quality control.[36] It was a strangely prophetic vision. Only in very recent years, with the adoption of artificial intelligence in the workplace, has Babbage's unusual twin goals of computation and worker automation become possible at scale.

Babbage's economic thought extended outward from Smith's but diverged in one important way. For Smith, the economic value of an object was understood in relation to the cost of the labor required to produce it. In Babbage's rendering, however, value in a factory was derived from investment in the design of the manufacturing process rather than from the labor force of its employees. The real innovation was the logistical process, while workers simply enacted the tasks defined for them and operated the machines as instructed.

For Babbage, labor's role in the value production chain was largely negative: workers might fail to perform their tasks in the timely manner prescribed by the precision machines they operated, whether through poor discipline, injury, absenteeism, or acts of resistance. As noted by historian Simon Schaffer, "Under Babbage's gaze, factories looked like perfect engines and calculating machines like perfect computers. The workforce might be a source of trouble—it could make tables err or factories fail—but it could not be seen as a source of value."[37] The factory is conceived as a rational calculating machine with only one weakness: its frail and untrustworthy human labor force.

Babbage's theory was, of course, heavily inflected with a kind of financial liberalism, causing him to view labor as a problem that needed to be contained by automation. There was little consideration of the human costs of this automation or of how automation might be put to use to improve the working lives of factory employees. Instead, Babbage's idealized machinery aimed primarily to maximize financial returns to the plant owners and their investors. In a similar vein, today's proponents of workplace AI present a vision of production that prioritizes efficiency, cost-cutting, and higher profits instead of, say, assisting their employees by replacing repetitive drudge work. As Astra Taylor argues, "The kind of efficiency to which techno-evangelists aspire emphasizes standardization, simplification, and speed, not diversity, complexity, and interdependence."[38] This should not surprise us: it is a necessary outcome of the standard business model of for-profit companies where the highest responsibility is to shareholder value. We are living the result of a system in which companies must extract as much value as possible. Meanwhile, 94 percent of all new American jobs created between 2005 and 2015 were for "alternative work"—jobs that fall outside of full-time, salaried employment.[39] As companies reap the benefits of increasing automation, people are, on average, working longer hours, in more jobs, for less pay, in insecure positions.

The Meat Market

Among the first industries to implement the type of mechanized production line Babbage envisioned was the Chicago
meat-packing industry in the 1870s. Trains brought livestock
to the stockyard gates; the animals were funneled toward their
slaughter in adjacent plants; and the carcasses were transported to various butchering and processing stations by means
of a mechanized overhead trolley system, forming what came
to be known as the *disassembly line.* The finished products
could be shipped to faraway markets in specially designed re-
frigerated rail cars.[40] Labor historian Harry Braverman noted
that the Chicago stockyards realized Babbage's vision of automation and division of labor so completely that the human
techniques required at any point on the disassembly line could
be performed by just about anyone.[41] Low-skill laborers could
be paid the bare minimum and replaced at the first sign of
trouble, themselves becoming as thoroughly commoditized as
the packaged meats they produced.

When Upton Sinclair wrote *The Jungle,* his harrow-
ing novel about working-class poverty, he set it in the meat-
packing plants of Chicago. Although his intended point was
to highlight the hardships of working immigrants in support
of a socialist political vision, the book had an entirely different
effect. The depictions of diseased and rotting meat prompted
a public outcry over food safety and resulted in the passing of
the Meat Inspection Act in 1906. But the focus on workers was
lost. Powerful institutions from the meat-packing industry to
Congress were prepared to intervene to improve the methods
of production, but addressing the more fundamental exploitative labor dynamics that propped up the entire system was off
limits. The persistence of this pattern underscores how power
responds to critique: whether the product is cow carcasses or

Armour Beef dressing floor, 1952.
Courtesy Chicago Historical Society

facial recognition, the response is to accept regulation at the margins but to leave untouched the underlying logics of production.

Two other figures loom large in the history of workplace automation: Henry Ford, whose moving assembly line from the early twentieth century was inspired by Chicago's disassembly lines, and Frederick Winslow Taylor, the founder of scientific management. Taylor forged his career in the latter years of the nineteenth century developing a systematic approach to workplace management, one that focused on the minute movements of workers' bodies. Whereas Smith's and Babbage's notion of the division of labor was intended to provide a way to distribute work between people and tools, Taylor narrowed his focus to include microscopic subdivisions in the actions of each worker.

As the latest technology for precisely tracking time, the stopwatch was to become a key instrument of workplace surveillance for shop-floor supervisors and production engineers alike. Taylor used stopwatches to perform studies of workers that included detailed breakdowns of the time taken to perform the discrete physical motions involved in any given task. His *Principles of Scientific Management* established a system to quantify the movements of workers' bodies, with a view to deriving an optimally efficient layout of tools and working processes. The aim was maximum output at minimal cost.[42] It exemplified Marx's description of the domination of clock time, "Time is everything, man is nothing; he is, at most, time's carcass."[43]

Foxconn, the largest electronics manufacturing company in the world, which makes Apple iPhones and iPads, is a vivid example of how workers are reduced to animal bodies performing tightly controlled tasks. Foxconn became notorious for its rigid and militaristic management protocols after a spate of suicides in 2010.[44] Just two years later, the company's chairman, Terry Gou, described his more than one million employees this way: "As human beings are also animals, to manage one million animals gives me a headache."[45]

Controlling time becomes another way to manage bodies. In service and fast-food industries, time is measured down to the second. Assembly line workers cooking burgers at McDonald's are assessed for meeting such targets as five seconds to process screen-based orders, twenty-two seconds to assemble a sandwich, and fourteen seconds to wrap the food.[46] Strict adherence to the clock removes margin for error from the system. The slightest delay (a customer taking too long to order, a coffee machine failing, an employee calling in sick) can result in a cascading ripple of delays, warning sounds, and management notifications.

Even before McDonald's workers join the assembly line, their time is being managed and tracked. An algorithmic scheduling system incorporating historical data analysis and demand-prediction models determines workers' shift allocations, resulting in work schedules that can vary from week to week and even day to day. A 2014 class action lawsuit against McDonald's restaurants in California noted that franchisees are led by software that gives algorithmic predictions regarding employee-to-sales ratios and instructs managers to reduce staff quickly when demand drops.[47] Employees reported being told to delay clocking in to their shifts and instead to hang around nearby, ready to return to work if the restaurant started getting busy again. Because employees are paid only for time clocked in, the suit alleged that this amounted to significant wage theft on the part of the company and its franchisees.[48]

Algorithmically determined time allocations will vary from extremely short shifts of an hour or less to very long ones during busy times—whatever is most profitable. The algorithm doesn't factor in the human costs of waiting or getting to work only to be sent home or being unable to predict one's schedule and plan one's life. This time theft helps the efficiency of the company, but it comes at the direct cost of the employees.

Managing Time, Privatizing Time

Fast-food entrepreneur Ray Kroc, who helped turn McDonald's into a global franchise, joined the lineage of Smith, Babbage, Taylor, and Ford when he designed the standard sandwich assembly line and made his employees follow it unthinkingly. Surveillance, standardization, and the reduction of individual craft were central to Kroc's method. As labor researchers Clare Mayhew and Michael Quinlan argue with regard to the McDonald's standardized process, "The Fordist management system documented work and production tasks in minuscule detail. It required on-going documented participation and entailed detailed control of each individual's work process. There was an almost total removal of all conceptual work from execution of tasks."[49]

Minimizing the time spent at each station, or cycle time, became an object of intense scrutiny within the Fordist factory, with engineers dividing work tasks into ever-smaller pieces so they could be optimized and automated, and with supervisors disciplining workers whenever they fell behind. Supervisors, even Henry Ford himself, could often be seen walking the length of the factory, stopwatch in hand, recording cycle times and noting any discrepancies in a station's productivity.[50]

Now employers can passively surveil their workforce without walking out onto the factory floor. Instead, workers clock in to their shifts by swiping access badges or by presenting their fingerprints to readers attached to electronic time clocks. They work in front of timing devices that indicate the minutes or seconds left to perform the current task before a manager is notified. They sit at workstations fitted with sensors that continuously report on their body temperature, their physical distance from colleagues, the amount of time they spend browsing websites instead of performing assigned tasks, and so on. WeWork, the coworking behemoth that burned itself out over the course of 2019, quietly fitted its work spaces with surveillance devices in an effort to create new forms of data monetization. Its 2019 acquisition of the spatial analytics startup Euclid raised concerns, with the suggestion that it planned to track its paying members as they moved through their facilities.[51] Domino's Pizza has added to its kitchens machine-vision systems that inspect a finished pizza to ensure the staff made it according to prescribed standards.[52] Surveillance apparatuses are justified for producing inputs for algorithmic scheduling systems that further modulate work time, or to glean behavioral signals that may correlate with signs of high or low performance, or merely sold to data brokers as a form of insight.

In her essay "How Silicon Valley Sets Time," sociology professor Judy Wajcman argues that the aims of time-tracking tools and the demographic makeup of Silicon Valley are no coincidence.[53] Silicon Valley's elite workforce "is even younger, more masculine and more fully committed to working all hours," while also creating productivity tools that are premised on a kind of ruthless, winner-takes-all race to maximal efficiency.[54] This means that young, mostly male engineers, often unencumbered by time-consuming familial or community responsibilities, are building the tools that will police very different workplaces, quantifying the productivity and desirability of employees. The workaholism and round-the-clock hours often glorified by tech start-ups become an implicit benchmark against which other workers are measured, producing a vision of a standard worker that is masculinized, narrow, and reliant on the unpaid or underpaid care work of others.

Private Time

The coordination of time has become ever more granular in the technological forms of workplace management. For example, General Motors' Manufacturing Automation Protocol (MAP) was an early attempt to provide standard solutions to common manufacturing robot coordination problems, including clock synchronization.[55] In due course, other, more generic time synchronization protocols that could be delivered over ethernet and TCP/IP networks emerged, including the Network Time Protocol (NTP), and, later, the Precision Time

Protocol (PTP), each of which spawned a variety of competing implementations across various operating systems. Both NTP and PTP function by establishing a hierarchy of clocks across a network, with a "master" clock driving the "slave" clocks.

The master-slave metaphor is riddled throughout engineering and computation. One of the earliest uses of this racist metaphor dates back to 1904 describing astronomical clocks in a Cape Town observatory.[56] But it wasn't until 1960s that the master-slave terminology spread, particularly after it was used in computing, starting with the Dartmouth timesharing system. Mathematicians John Kemeny and Thomas Kurtz developed a time-sharing program for access to computing resources after a suggestion by one of the early founders of AI, John McCarthy. As they wrote in *Science* in 1968, "First, all computing for users takes place in the slave computer, while the executive program (the 'brains' of the system) resides in the master computer. It is thus impossible for an erroneous or runaway user program in the slave computer to 'damage' the executive program and thereby bring the whole system to a halt."[57] The problematic implication that control is equivalent to intelligence would continue to shape the AI field for decades. And as Ron Eglash has argued, the phrasing has a strong echo of the pre–Civil War discourse on runaway slaves.[58]

The master-slave terminology has been seen as offensive by many and has been removed from Python, a coding language common in machine learning, and Github, a software development platform. But it persists in one of the most expansive computational infrastructures in the world. Google's Spanner—named as such because it spans the entire planet—is a massive, globally distributed, synchronously replicated database. It is the infrastructure that supports Gmail, Google search, advertising, and all of Google's distributed services.

At this scale, functioning across the globe, Spanner synchronizes time across millions of servers in hundreds of data centers. Every data center has a "time master" unit that is always receiving GPS time. But because servers were polling a variety of master clocks, there was slight network latency and clock drift. How to resolve this uncertainty? The answer was to create a new distributed time protocol—a proprietary form of time—so that all servers could be in sync regardless of where they were across the planet. Google called this new protocol, without irony, TrueTime.

Google's TrueTime is a distributed time protocol that functions by establishing trust relationships between the local clocks of data centers so they can decide which peers to synchronize with. Benefiting from a sufficiently large number of reliable clocks, including GPS receivers and atomic clocks that provide an extremely high degree of precision, and from sufficiently low levels of network latency, TrueTime allows a distributed set of servers to guarantee that events can occur in a determinate sequence across a wide area network.[59]

What's most remarkable in this system of privatized Google time is how TrueTime manages uncertainty when there is clock drift on individual servers. "If the uncertainty is large, Spanner slows down to wait out that uncertainty," Google researchers explain.[60] This embodies the fantasy of slowing down time, of moving it at will, and of bringing the planet under a single proprietary time code. If we think of the human experience of time as something shifting and subjective, moving faster or slower depending on where we are and whom we are with, then this is a social experience of time. TrueTime is the ability to create a shifting timescale under the control of a centralized master clock. Just as Isaac Newton imagined an absolute form of time that exists independently of any perceiver, Google has invented its own form of universal time.

Proprietary forms of time have long been used to make machines run smoothly. Railroad magnates in the nineteenth century had their own forms of time. In New England in 1849, for example, all trains were to adopt "true time at Boston as given by William Bond & Son, No. 26 Congress Street."[61] As Peter Galison has documented, railroad executives weren't fond of having to switch to other times depending on which state their trains traveled to, and the general manager of the New York & New England Railroad Company called switching to other times "a nuisance and great inconvenience and no use to anybody I can see."[62] But after a head-on train collision killed fourteen people in 1853, there was immense pressure to coordinate all of the clocks using the new technology of the telegraph.

Like artificial intelligence, the telegraph was hailed as a unifying technology that would expand the capabilities of human beings. In 1889 Lord Salisbury boasted that the telegraph had "assembled all mankind upon one great plane."[63] Businesses, governments, and the military used the telegraph to compile time into a coherent grid, erasing more local forms of timekeeping. And the telegraph was dominated by one of the first great industrial monopolies, Western Union. In addition to altering the temporal and spatial boundaries of human interaction, communications theorist James Carey argues that the telegraph also enabled a new form of monopoly capitalism: "a new body of law, economic theory, political arrangements, management techniques, organizational structures, and scientific rationales with which to justify and make effective the development of a privately owned and controlled monopolistic corporation."[64] While this interpretation implies a kind of technological determinism in what was a complex series of developments, it is fair to say that the telegraph—paired with the transatlantic cable—enabled imperial powers to maintain more centralized control over their colonies.

The telegraph made time a central focus for commerce. Rather than traders exploiting the difference in prices between regions by buying low and selling high in varying locations, now they traded between time zones: in Carey's terms, a shift from space to time, from arbitrage to futures.[65] The privatized time zones of data centers are just the latest example. The infrastructural ordering of time acts as a kind of "macrophysics of power," determining new logics of information at a planetary level.[66] Such power is necessarily centralizing, creating orders of meaning that are extremely difficult to see, let alone disrupt.

Defiance of centralized time is a vital part of this history. In the 1930s, when Ford wanted more control over his global supply chain, he set up a rubber plantation and processing facility deep in the Brazilian rain forest, in a town he named Fordlandia. He employed local workers to process rubber for shipping back to Detroit, but his attempts to impose his tightly controlled manufacturing process on the local population backfired. Rioting workers tore apart the factory's time clocks, smashing the devices used to track the entry and exit of each worker in the plant.

Other forms of insurgence have centered on adding friction to the work process. The French anarchist Émile Pouget used the term "sabotage" to mean the equivalent of a "go slow" on the factory floor, when workers intentionally reduce their pace of work.[67] The objective was to withdraw efficiency, to reduce the value of time as a currency. Although there will always be ways to resist the imposed temporality of work, with forms of algorithmic and video monitoring, this becomes much harder—as the relation between work and time is observed at ever closer range.

From the fine modulations of time within factories to the big modulations of time at the scale of planetary computation networks, defining time is an established strategy for centralizing power. Artificial intelligence systems have allowed for greater exploitation of distributed labor around the world to take advantage of uneven economic topologies. Simultaneously, the tech sector is creating for itself a smooth global terrain of time to strengthen and speed its business objectives. Controlling time—whether via the clocks for churches, trains, or data centers—has always been a function of controlling the political order. But this battle for control has never been smooth, and it is a far-reaching conflict. Workers have found ways to intervene and resist, even when technological developments were forced on them or presented as desirable improvements, particularly if the only refinements were to increase surveillance and company control.

Setting the Rate

Amazon goes to great lengths to control what members of the public can see when they enter a fulfillment center. We are told about the fifteen-dollar-an-hour minimum wage and the perks for employees who can last longer than a year, and we are shown brightly lit break rooms that have Orwellian corporate slogans painted on the walls: "Frugality," "Earn trust of others," and "Bias for action." The official Amazon guide cheerily explains what is happening at predetermined stops with rehearsed vignettes. Any questions about labor conditions are carefully answered to paint the most positive picture. But there are signs of unhappiness and dysfunction that are much harder to manage.

Out on the picking floor, where associates must pick up gray containers (known as "totes") full of purchases to ship,

Fordlandia Time Clock, destroyed in the riot of December 1930.
From the Collections of The Henry Ford

whiteboards bear the marks of recent meetings. One had multiple complaints that the totes were stacked too high and that constantly reaching up to grab them was causing considerable pain and injuries. When asked about this, the Amazon guide quickly responded that this concern was being addressed by lowering the height of the conveyor belt in key sections. This was seen as a success: a complaint had been registered and action would be taken. The guide took this opportunity to explain for the second time that this was why unions were unnecessary here, because "associates have many opportunities to interface with their managers," and unionization only interferes with communication.[68]

But on the way out of the facility, I walked past a live feed of messages from workers on a large flat screen, with a sign

above it that read, "The Voice of the Associates." This was far
less varnished. Messages scrolled rapidly past with complaints
about arbitrary scheduling changes, the inability to book vacation time near holidays, and missing family occasions and
birthdays. Pat responses from management seemed to be multiple variations on the theme of "We value your feedback."

"Enough is enough. Amazon, we want you to treat us like
humans, and not like robots."[69] These are the words of Abdi
Muse, executive director of the Awood Center in Minneapolis, a community organization that advocates for the working
conditions of Minnesota's East African populations. Muse is a
soft-spoken defender of Amazon warehouse workers who are
pushing for better working conditions. Many workers in his
Minnesota community have been hired by Amazon, which actively recruited them and added sweeteners to the deal, such as
free busing to work.

What Amazon didn't advertise was "the rate"—the
worker productivity metric driving the fulfillment centers that
quickly became unsustainable and, according to Muse, inhumane. Workers began suffering high stress, injuries, and illness. Muse explained that if their rate went down three times
they would be fired, no matter how long they had worked at
the warehouse. Workers talked about having to skip bathroom
breaks for fear that they would underperform.

But the day we met, Muse was optimistic. Even though
Amazon explicitly discourages unions, informal groups of
workers were springing up across the United States and staging
protests. He smiled widely as he reported that the organizing
was starting to have an impact. "Something incredible is happening," he told me. "Tomorrow a group of Amazon workers
will be walking off the job. It's such a courageous group of
women, and they are the real heroes."[70] Indeed, that night, approximately sixty warehouse workers walked out of a deliv-

ery center in Eagan, Minnesota, wearing their mandated yellow vests. They were mostly women of Somali descent, and they held up signs in the rain, demanding such improvements as increased wages for night shifts and weight restrictions on boxes.[71] Only a few days earlier, Amazon workers in Sacramento, California, had protested the firing of an employee who had gone one hour over her bereavement leave after a family member died. Two weeks before that, more than a thousand Amazon workers staged the first ever white-collar walkout in the company's history over its massive carbon footprint.

Eventually, Amazon's representatives in Minnesota came to the table. They were happy to discuss many issues but never "the rate." "They said forget about 'the rate,'" recounted Muse. "We can talk about other issues, but the rate is our business model. We cannot change that."[72] The workers threatened to walk away from the table, and still Amazon would not budge. For both sides, "the rate" was the core issue, but it was also the hardest to alter. Unlike other local labor disputes where the on-the-ground supervisors might have been able to make concessions, the rate was set based on what the executives and tech workers in Seattle—far removed from the warehouse floor—had decided and had programmed Amazon's computational distribution infrastructure to optimize for. If the local warehouses were out of sync, Amazon's ordering of time was threatened. Workers and organizers started to see this as the real issue. They are shifting their focus accordingly toward building a movement across different factories and sectors of Amazon's workforce to address the core issues of power and centralization represented by the relentless rhythm of "the rate" itself.

These fights for time sovereignty, as we've seen, have a history. AI and algorithmic monitoring are simply the latest technologies in the long historical development of factories, timepieces, and surveillance architectures. Now many more sectors—from Uber drivers to Amazon warehouse workers to highly paid Google engineers—perceive themselves in this shared fight. This was strongly articulated by the executive director of the New York Taxi Workers Alliance, Bhairavi Desai, who put it this way: "Workers always know. They are out there building solidarity with each other, at red lights or in restaurants or in hotel queues, because they know that in order to prosper they have to band together."[73] Technologically driven forms of worker exploitation are a widespread problem in many industries. Workers are fighting against the logics of production and the order of time they must work within. The structures of time are never completely inhumane, but they are maintained right at the outer limit of what most people can tolerate.

Cross-sector solidarity in labor organizing is nothing new. Many movements, such as those led by traditional labor unions, have connected workers in different industries to win the victories of paid overtime, workplace safety, parental leave, and weekends. But as powerful business lobbies and neoliberal governments have chipped away at labor rights and protections over the past several decades and limited the avenues for worker organizing and communications, cross-sector support has become more difficult.[74] Now AI-driven systems of extraction and surveillance have become a shared locus for labor organizers to fight as a unified front.[75]

"We are all tech workers" has become a common sign at tech-related protests, carried by programmers, janitors, cafeteria workers, and engineers alike.[76] It can be read in multiple ways: it demands that the tech sector recognize the wide labor force it draws on to make its products, infrastructures, and workplaces function. It also reminds us that so many workers use laptops and mobile devices for work, engage on platforms like Facebook or Slack, and are subject to forms of workplace

AI systems for standardization, tracking, and assessment. This has set the stage for a form of solidarity built around tech work. But there are risks in centering tech workers and technology in what are more generalized and long-standing labor struggles. All kinds of workers are subject to the extractive technical infrastructures that seek to control and analyze time to its finest grain — many of whom have no identification with the technology sector or tech work at all. The histories of labor and automation remind us that what is at stake is producing more just conditions for every worker, and this broader goal should not depend on expanding the definition of tech work in order to gain legitimacy. We all have a collective stake in what the future of work looks like.

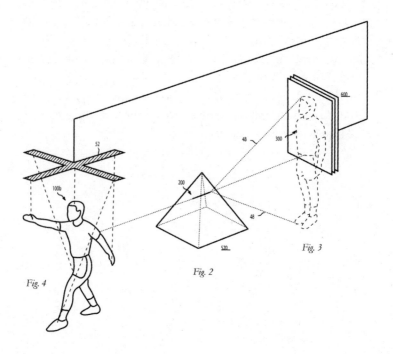

52

100b

Fig. 4

200

520

Fig. 2

48

300

48

600

Fig. 3

3
Data

A young woman gazes upward, eyes focused on something outside the frame, as though she is refusing to acknowledge the camera. In the next photograph, her eyes are locked on the middle distance. Another image shows her with disheveled hair and a downcast expression. Over the sequence of photos we see her aging over time, and the lines around her mouth turn down and deepen. In the final frame she appears injured and dispirited. These are mug shots of a woman across multiple arrests over many years of her life. Her images are contained in a collection known as NIST Special Database 32–Multiple Encounter Dataset, which is shared on the internet for researchers who would like to test their facial recognition software.[1]

This dataset is one of several maintained by the National Institute of Standards and Technology (NIST), one of the oldest and most respected physical science laboratories in the United States and now part of the Department of Commerce. NIST was created in 1901 to bolster the nation's measurement infrastructure and to create standards that could compete with economic rivals in the industrialized world, such as Germany

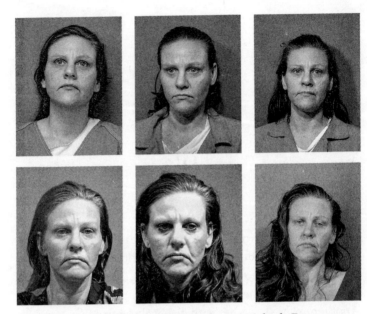

Images from NIST Special Database 32—Multiple Encounter
Dataset (MEDS). National Institute of Standards and Technology,
U.S. Department of Commerce

and the United Kingdom. Everything from electronic health
records to earthquake-resistant skyscrapers to atomic clocks
is under the purview of NIST. It became the agency of measurement: of time, of communications protocols, of inorganic
crystal structures, of nanotechnology.[2] NIST's purpose is to
make systems interoperable through defining and supporting
standards, and this now includes developing standards for artificial intelligence. One of the testing infrastructures it main-
tains is for biometric data.

I first discovered the mug shot databases in 2017 when I
was researching NIST's data archives. Their biometric collections are extensive. For more than fifty years, NIST has col-
laborated with the Federal Bureau of Investigation on auto-

mated fingerprint recognition and has developed methods to assess the quality of fingerprint scanners and imaging systems.[3] After the terrorist attacks of September 11, 2001, NIST became part of the national response to create biometric standards to verify and track people entering the United States.[4] This was a turning point for research on facial recognition; it widened out from a focus on law enforcement to controlling people crossing national borders.[5]

The mug shot images themselves are devastating. Some people have visible wounds, bruising, and black eyes; some are distressed and crying. Others stare blankly back at the camera. Special Dataset 32 contains thousands of photographs of deceased people with multiple arrests, as they endured repeated encounters with the criminal justice system. The people in the mug shot datasets are presented as data points; there are no stories, contexts, or names. Because mug shots are taken at the time of arrest, it's not clear if these people were charged, acquitted, or imprisoned. They are all presented alike.

The inclusion of these images in the NIST database has shifted their meaning from being used to identify individuals in systems of law enforcement to becoming the technical baseline to test commercial and academic AI systems for detecting faces. In his account of police photography, Allan Sekula has argued that mug shots are part of a tradition of technical realism that aimed to "provide a standard physiognomic gauge of the criminal."[6] There are two distinct approaches in the history of the police photograph, Sekula observes. Criminologists like Alphonse Bertillon, who invented the mug shot, saw it as a kind of biographical machine of identification, necessary to spot repeat offenders. On the other hand, Francis Galton, the statistician and founding figure of eugenics, used composite portraiture of prisoners as a way to detect a biologically determined "criminal type."[7] Galton was working within a physiognomist paradigm in which the goal was to find a generalized look that could be used to identify deep character traits from external appearances. When mug shots are used as training data, they function no longer as tools of identification but rather to fine-tune an automated form of vision. We might think of this as Galtonian formalism. They are used to detect the basic mathematical components of faces, to "reduce nature to its geometrical essence."[8]

Mug shots form part of the archive that is used to test facial-recognition algorithms. The faces in the Multiple Encounter Dataset have become standardized images, a technical substrate for comparing algorithmic accuracy. NIST, in collaboration with the Intelligence Advanced Research Projects Activity (IARPA), has run competitions with these mug shots in which researchers compete to see whose algorithm is the fastest and most accurate. Teams strive to beat one another at tasks like verifying the identity of faces or retrieving a face from a frame of surveillance video.[9] The winners celebrate these victories; they can bring fame, job offers, and industrywide recognition.[10]

Neither the people depicted in the photographs nor their families have any say about how these images are used and likely have no idea that they are part of the test beds of AI. The subjects of the mug shots are rarely considered, and few engineers will ever look at them closely. As the NIST document describes them, they exist purely to "refine tools, techniques, and procedures for face recognition as it supports Next Generation Identification (NGI), forensic comparison, training, analysis, and face image conformance and inter-agency exchange standards."[11] The Multiple Encounter Dataset description observes that many people show signs of enduring violence, such as scars, bruises, and bandages. But the document concludes that these signs are "difficult to interpret due to the lack of ground truth for comparison with a 'clean' sample."[12] These people are not seen so much as individuals but as part of a shared technical resource—just another data component of the Facial Recognition Verification Testing program, the gold standard for the field.

I've looked at hundreds of datasets over years of research into how AI systems are built, but the NIST mug shot databases are particularly disturbing because they represent the model of what was to come. It's not just the overwhelming pathos of the images themselves. Nor is it solely the invasion of privacy they represent, since suspects and prisoners have no right to refuse being photographed. It's that the NIST databases foreshadow the emergence of a logic that has now thoroughly pervaded the tech sector: the unswerving belief that everything is data and is there for the taking. It doesn't matter where a photograph was taken or whether it reflects a moment of vulnerability or pain or if it represents a form of shaming the subject. It has become so normalized across the industry to take and use whatever is available that few stop to question the underlying politics.

Mug shots, in this sense, are the urtext of the current approach to making AI. The context—and exertion of power—that these images represent is considered irrelevant because they no longer exist as distinct things unto themselves. They are not seen to carry meanings or ethical weight as images of individual people or as representations of structural power in the carceral system. The personal, the social, and the political meanings are all imagined to be neutralized. I argue this represents a shift from *image* to *infrastructure*, where the meaning or care that might be given to the image of an individual person, or the context behind a scene, is presumed to be erased at the moment it becomes part of an aggregate mass that will drive a broader system. It is all treated as data to be run through functions, material to be ingested to improve technical performance. This is a core premise in the ideology of data extraction.

Machine learning systems are trained on images like these every day—images that were taken from the internet or from state institutions without context and without consent. They are anything but neutral. They represent personal histories, structural inequities, and all the injustices that have accompanied the legacies of policing and prison systems in the United States. But the presumption that somehow these images can serve as apolitical, inert material influences how and what a machine learning tool "sees." A computer vision system can detect a face or a building but not why a person was inside a police station or any of the social and historical context surrounding that moment. Ultimately, the specific instances of data—a picture of a face, for example—aren't considered to matter for training an AI model. All that matters is a sufficiently varied aggregate. Any individual image could easily be substituted for another and the system would work the same. According to this worldview, there is always more data to capture from the constantly growing and globally distributed treasure chest of the internet and social media platforms.

A person standing in front of a camera in an orange jumpsuit, then, is dehumanized as just more data. The history of these images, how they were acquired, and their institutional, personal, and political contexts are not considered relevant. The mug shot collections are used like any other practical resource of free, well-lit images of faces, a benchmark to make tools like facial recognition function. And like a tightening ratchet, the faces of deceased persons, suspects, and prisoners are harvested to sharpen the police and border surveillance facial recognition systems that are then used to monitor and detain more people.

The last decade has seen a dramatic capture of digital material for AI production. This data is the basis for sense-making in AI, not as classical representations of the world with individual meaning, but as a mass collection of data for machine abstractions and operations. This large-scale capture has become so fundamental to the AI field that it is unquestioned. So how did we get here? What ways of conceiving data have facilitated this stripping of context, meaning, and specificity? How is training data acquired, understood, and used in machine learning? In what ways does training data limit *what* and *how* AI interprets the world? What forms of power do these approaches enhance and enable?

In this chapter I show how data has become a driving force in the success of AI and its mythos and how everything that can be readily captured is being acquired. But the deeper implications of this standard approach are rarely addressed, even as it propels further asymmetries of power. The AI industry has fostered a kind of ruthless pragmatism, with minimal context, caution, or consent-driven data practices while promoting the idea that the mass harvesting of data is necessary and justified for creating systems of profitable computational "intelligence." This has resulted in a profound metamorphosis, where all forms of image, text, sound, and video are just raw data for AI systems and the ends are thought to justify the means. But we should ask: Who has benefited most from this transformation, and why have these dominant narratives of data persisted? And as we saw in the previous chapters, the logic of extraction that has shaped the relationship to the earth and to human labor is also a defining feature of how data is used and understood in AI. By looking closely at training data as a central example in the ensemble of machine learning, we can begin to see what is at stake in this transformation.

Training Machines to See

It's useful to consider why machine learning systems currently demand massive amounts of data. One example of the problem in action is computer vision, the subfield of artificial intelligence concerned with teaching machines to detect and interpret images. For reasons that are rarely acknowledged in the field of computer science, the project of interpreting images is a profoundly complex and relational endeavor. Images are remarkably slippery things, laden with multiple potential meanings, irresolvable questions, and contradictions. Yet now it's common practice for the first steps of creating a computer vision system to scrape thousands—or even millions—of images from the internet, create and order them into a series of classifications, and use this as a foundation for how the system will perceive observable reality. These vast collections are called training datasets, and they constitute what AI developers often refer to as "ground truth."[13] Truth, then, is less about a factual representation or an agreed-upon reality and more commonly about a jumble of images scraped from whatever various online sources were available.

For supervised machine learning, human engineers supply labeled training data to a computer. Two distinct types of algorithms then come into play: *learners* and *classifiers*. The learner is the algorithm that is trained on these labeled data examples; it then informs the classifier how best to analyze the relation between the new inputs and the desired target output (or prediction). It might be predicting whether a face is contained in an image or whether an email is spam. The more examples of correctly labeled data there are, the better the algorithm will be at producing accurate predictions. There are many kinds of machine learning models, including neural networks, logistic regression, and decision trees. Engineers will choose a model based on what they are building—be it a facial recognition system or a means of detecting sentiment on social media—and fit it to their computational resources.

Consider the task of building a machine learning system that can detect the difference between pictures of apples and oranges. First, a developer has to collect, label, and train a neural network on thousands of labeled images of apples and oranges. On the software side, the algorithms conduct a statistical survey of the images and develop a model to recognize the difference between the two classes. If all goes according to plan, the trained model will be able to distinguish the difference between images of apples and oranges that it has never encountered before.

But if, in our example, all of the training images of apples are red and none are green, then a machine learning system might deduce that "all apples are red." This is what is known as an *inductive inference,* an open hypothesis based on available data, rather than a *deductive inference,* which follows logically from a premise.[14] Given how this system was trained, a green apple wouldn't be recognized as an apple at all. Training datasets, then, are at the core of how most machine learning systems make inferences. They serve as the primary source material that AI systems use to form the basis of their predictions.

Training data also defines more than just the features of machine learning algorithms. It is used to assess how they perform over time. Like prized thoroughbreds, machine learning algorithms are constantly raced against one another in competitions all over the world to see which ones perform the best with a given dataset. These benchmark datasets become the alphabet on which a *lingua franca* is based, with many labs from multiple countries converging around canonical sets to try to outperform one another. One of the best-known competitions is the ImageNet Challenge, where researchers compete to see whose methods can most accurately classify and detect objects and scenes.[15]

Once training sets have been established as useful benchmarks, they are commonly adapted, built upon, and expanded. As we will see in the next chapter, a type of genealogy of training sets emerges—they inherit learned logic from earlier examples and then give rise to subsequent ones. For example, ImageNet draws on the taxonomy of words inherited from the influential 1980s lexical database known as WordNet; and WordNet inherits from many sources, including the Brown Corpus of one million words, published in 1961. Training datasets stand on the shoulders of older classifications and collections. Like an expanding encyclopedia, the older forms remain and new items are added over decades.

Training data, then, is the foundation on which contemporary machine learning systems are built.[16] These datasets shape the epistemic boundaries governing how AI operates and, in that sense, create the limits of how AI can "see" the world. But training data is a brittle form of ground truth—and even the largest troves of data cannot escape the fundamental slippages that occur when an infinitely complex world is simplified and sliced into categories.

A Brief History of the Demand for Data

"The world has arrived at an age of cheap complex devices of great reliability; and something is bound to come of it." So said Vannevar Bush, the inventor and administrator who oversaw the Manhattan Project as director of the Office of Scientific Research and Development and later was integral to the creation of the National Science Foundation. It was July 1945; the bombs were yet to drop on Hiroshima and Nagasaki, and Bush had a theory about a new kind of data-connecting system that was yet to be born. He envisaged the "advanced arithmetical machines of the future" that would perform at extremely fast speed and "select their own data and manipulate it in accordance with the instructions." But the machines would need monumental amounts of data: "Such machines will have enormous appetites. One of them will take instructions and data from a whole roomful of girls armed with simple key board punches, and will deliver sheets of computed results every few minutes. There will always be plenty of things to compute in the detailed affairs of millions of people doing complicated things."[17]

The "roomful of girls" Bush referred to were the keypunch operators doing the day-to-day work of computation. As historians Jennifer Light and Mar Hicks have shown, these women were often dismissed as input devices for intelligible data records. In fact, their role was just as important to crafting data and making systems work as that of the engineers who designed the wartime-era digital computers.[18] But the relationship between data and processing machinery was already being imagined as one of endless consumption. The machines would be data-hungry, and there would surely be a wide horizon of material to extract from millions of people.

In the 1970s, artificial intelligence researchers were mainly exploring what's called an expert systems approach: rules-based programming that aims to reduce the field of possible actions by articulating forms of logical reasoning. But it quickly became evident that this approach was fragile and impractical in real-world settings, where a rule set was rarely able to handle uncertainty and complexity.[19] New approaches were needed. By the mid-1980s, research labs were turning toward probabilistic or brute force approaches. In short, they were using lots of computing cycles to calculate as many options as possible to find the optimal result.

One significant example was the speech recognition group at IBM Research. The problem of speech recognition had primarily been dealt with using linguistic methods, but then information theorists Fred Jelinek and Lalit Bahl formed a new group, which included Peter Brown and Robert Mercer (long before Mercer became a billionaire, associated with funding Cambridge Analytica, Breitbart News, and Donald Trump's 2016 presidential campaign). They tried something different. Their techniques ultimately produced precursors for the speech recognition systems underlying Siri and Dragon Dictate, as well as machine translation systems like Google Translate and Microsoft Translator.

They started using statistical methods that focused more on how often words appeared in relation to one another, rather than trying to teach computers a rules-based approach using grammatical principles or linguistic features. Making this statistical approach work required an enormous amount of real speech and text data, or training data. The result, as media scholar Xiaochang Li writes, was that it required "a radical reduction of speech to merely data, which could be modeled and interpreted in the absence of linguistic knowledge or understanding. Speech *as such* ceased to matter." This shift was incredibly significant, and it would become a pattern repeated for decades: the reduction from context to data, from meaning to statistical pattern recognition. Li explains:

> The reliance on data over linguistic principles, however, presented a new set of challenges, for it meant that the statistical models were necessarily determined by the characteristics of training data. As a result, the size of the dataset became a central concern. . . . Larger datasets of observed out-

comes not only improved the probability estimates for a random process, but also increased the chance that the data would capture more rarely-occurring outcomes. Training data size, in fact, was so central to IBM's approach that in 1985, Robert Mercer explained the group's outlook by simply proclaiming, "There's no data like more data."[20]

For several decades, that data was remarkably hard to come by. As Lalit Bahl describes in an interview with Li, "Back in those days . . . you couldn't even find a million words in computer-readable text very easily. And we looked all over the place for text."[21] They tried IBM technical manuals, children's novels, patents of laser technology, books for the blind, and even the typed correspondence of IBM Fellow Dick Garwin, who created the first hydrogen bomb design.[22] Their method strangely echoed a short story by the science fiction author Stanislaw Lem, in which a man called Trurl decides to build a machine that would write poetry. He starts with "eight hundred and twenty tons of books on cybernetics and twelve thousand tons of the finest poetry."[23] But Trurl realizes that to program an autonomous poetry machine, one needs "to repeat the entire Universe from the beginning—or at least a good piece of it."[24]

Ultimately, the IBM Continuous Speech Recognition group found their "good piece" of the universe from an unlikely source. A major federal antitrust lawsuit was filed against IBM in 1969; the proceedings lasted for thirteen years, and almost a thousand witnesses were called. IBM employed a large staff just to digitize all of the deposition transcripts onto Hollerith punch cards. This ended up creating a corpus of a hundred million words by the mid-1980s. The notoriously antigovernment Mercer called this a "case of utility accidentally created by the government in spite of itself."[25]

IBM wasn't the only group starting to gather words by the ton. From 1989 to 1992, a team of linguists and computer scientists at the University of Pennsylvania worked on the Penn Treebank Project, an annotated database of text. They collected four and a half million words of American English for the purpose of training natural language processing systems. Their sources included Department of Energy abstracts, Dow Jones newswire articles, and Federal News Service reports of "terrorist activity" in South America.[26] The emerging text collections borrowed from earlier collections and then contributed new sources. Genealogies of data collections began to emerge, each building on the last—and often importing the same peculiarities, issues, or omissions wholesale.

Another classic corpus of text came from the fraud investigations of Enron Corporation after it declared the largest bankruptcy in American history. The Federal Energy Regulatory Commission seized the emails of 158 employees for the purposes of legal discovery.[27] It also decided to release these emails online because "the public's right to disclosure outweighs the individual's right to privacy."[28] This became an extraordinary collection. Over half a million exchanges in everyday speech could now be used as a linguistic mine: one that nonetheless represented the gender, race, and professional skews of those 158 workers. The Enron corpus has been cited in thousands of academic papers. Despite its popularity, it is rarely looked at closely: the *New Yorker* described it as "a canonic research text that no one has actually read."[29] This construction of and reliance on training data anticipated a new way of doing things. It transformed the field of natural language processing and laid the foundations of what would become normal practice in machine learning.

The seeds of later problems were planted here. Text archives were seen as neutral collections of language, as though there was a general equivalence between the words in a technical manual and how people write to colleagues via email. All text was repurposable and swappable, so long as there was enough of it that it could train a language model to predict with high levels of success what word might follow another. Like images, text corpuses work on the assumption that all training data is interchangeable. But language isn't an inert substance that works the same way regardless of where it is found. Sentences taken from Reddit will be different from those composed by executives at Enron. Skews, gaps, and biases in the collected text are built into the bigger system, and if a language model is based on the kinds of words that are clustered together, it matters where those words come from. There is no neutral ground for language, and all text collections are also accounts of time, place, culture, and politics. Further, languages that have less available data are not served by these approaches and so are often left behind.[30]

Clearly there are many histories and contexts that combine within IBM's training data, the Enron archive, or the Penn Treebank. How do we unpack what is and is not meaningful to understand these datasets? How does one communicate warnings like, "This dataset likely reflects skews related to its reliance on news stories about South American terrorists in the 1980s"? The origins of the underlying data in a system can be incredibly significant, and yet there are still, thirty years later, no standardized practices to note where all this data came from or how it was acquired—let alone what biases or classificatory politics these datasets contain that will influence all the systems that come to rely on them.[31]

Capturing the Face

While computer-readable text was becoming highly valued for speech recognition, the human face was the core concern for building systems of facial recognition. One central example emerged in the last decade of the twentieth century, funded by the Department of Defense CounterDrug Technology Development Program Office. It sponsored the Face Recognition Technology (FERET) program to develop automatic face recognition for intelligence and law enforcement. Before FERET, little training data of human faces was available, only a few collections of fifty or so faces here and there—not enough to do facial recognition at scale. The U.S. Army Research Laboratory led the technical project of creating a training set of portraits of more than a thousand people, in multiple poses, to make a grand total of 14,126 images. Like NIST's mug shot collections, FERET became a standard benchmark—a shared measuring tool to compare approaches for detecting faces.

The tasks that the FERET infrastructure was created to support included, once again, automated searching of mug shots, as well as monitoring airports and border crossings and searching driver's license databases for "fraud detection" (multiple welfare claims was a particular example mentioned in FERET research papers).[32] But there were two primary testing scenarios. In the first, an electronic mug book of known individuals would be presented to an algorithm, which then had to locate the closest matches from a large gallery. The second scenario focused on border and airport control: identifying a known individual—"smugglers, terrorists, or other criminals"—from a large population of unknown people.

These photographs are machine-readable by design, and not meant for human eyes, yet they make for remarkable viewing. The images are surprisingly beautiful—high-resolution photographs captured in the style of formal portraiture. Taken with 35 mm cameras at George Mason University, the tightly framed headshots depict a wide range of people, some of whom seem to have dressed for the occasion with carefully styled hair, jewelry, and makeup. The first set of photographs, taken between 1993 and 1994, are like a time capsule of early nineties haircuts and fashion. The subjects were asked to turn their heads to multiple positions; flicking through the images, you can see profile shots, frontal images, varying levels of illumination, and sometimes different outfits. Some subjects were photographed over several years, in order to begin to study how to track people as they age. Each subject was briefed about the project and signed a release form that had been approved by the university's ethics review board. Subjects knew what they were participating in and gave full consent.[33] This level of consent would become a rarity in later years.

FERET was the high-water mark of a formal style of "making data," before the internet began offering mass extraction without any permissions or careful camera work. Even at this early stage, though, there were problems with the lack of diversity of the faces collected. The FERET research paper from 1996 admits that "some questions were raised about the age, racial, and sexual distribution of the database" but that "at this stage of the program, the key issue was algorithm performance on a database of a large number of individuals."[34] Indeed, FERET was extraordinarily useful for this. As the interest in terrorist detection intensified and funding for facial recognition dramatically increased after 9/11, FERET became the most commonly used benchmark. From that point onward, biometric tracking and automated vision systems would rapidly expand in scale and ambition.

From the Internet to ImageNet

The internet, in so many ways, changed everything; it came to be seen in the AI research field as something akin to a natural resource, there for the taking. As more people began to upload their images to websites, to photo-sharing services, and ultimately to social media platforms, the pillaging began in earnest. Suddenly, training sets could reach a size that scientists in the 1980s could never have imagined. Gone was the need to stage photo shoots using multiple lighting conditions, controlled parameters, and devices to position the face. Now there were millions of selfies in every possible lighting condition, position, and depth of field. People began to share their baby photos, family snaps, and images of how they looked a decade ago, an ideal resource for tracking genetic similarity and face aging. Trillions of lines of text, containing both formal and informal forms of speech, were published every day. It was all grist for the mills of machine learning. And it was vast. As an example, on an average day in 2019, approximately 350 million photographs were uploaded to Facebook and 500 million tweets were sent.[35] And that's just two platforms based in the United States. Anything and everything online was primed to become a training set for AI.

The tech industry titans were now in a powerful position: they had a pipeline of endlessly refreshing images and text, and the more people shared their content, the more the tech industry's power grew. People would happily label their photographs with names and locations, free of charge, and that unpaid labor resulted in having more accurate, labeled data for machine vision and language models. Within the industry, these collections are highly valuable. They are proprietary troves that are rarely shared, given both the privacy issues and the competitive advantage they represent. But those outside the industry, such as the leading computer science labs in academia, wanted the same advantages. How could they afford to harvest people's data and have it hand-labeled by willing human participants? That's when new ideas began to emerge: combining images and text extracted from the internet with the labor of low-paid crowdworkers.

One of the most significant training sets in AI is ImageNet. It was first conceptualized in 2006, when Professor Fei-Fei Li decided to build an enormous dataset for object recognition. "We decided we wanted to do something that was completely historically unprecedented," Li said. "We're going to map out the entire world of objects."[36] The breakthrough research poster was published by the ImageNet team at a computer vision conference in 2009. It opened with this description:

> The digital era has brought with it an enormous explosion of data. The latest estimations put a number of more than 3 billion photos on Flickr, a similar number of video clips on YouTube and an even larger number for images in the Google Image Search database. More sophisticated and robust models and algorithms can be proposed by exploiting these images, resulting in better applications for users to index, retrieve, organize and interact with these data.[37]

From the outset, data was characterized as something voluminous, disorganized, impersonal, and ready to be exploited. According to the authors, "Exactly how such data can be utilized and organized is a problem yet to be solved." By extracting millions of images from the internet, primarily from search engines using the image-search option, the team produced a "large-scale ontology of images" that was meant to serve as a resource for "providing critical training and bench-marking data" for object and image recognition algorithms. Using this approach, ImageNet grew enormous. The team mass-harvested more than fourteen million images from the internet to be organized into more than twenty thousand categories. Ethical concerns about taking people's data were not mentioned in any of the team's research papers, even though many thousands of the images were of a highly personal and compromising nature.

Once the images had been scraped from the internet, a major concern arose: Who would label them all and put them into intelligible categories? As Li describes it, the team's first plan was to hire undergraduate students for ten dollars an hour to find images manually and add them to the dataset.[38] But she realized that with their budget, it would take more than ninety years to complete the project. The answer came when a student told Li about a new service: Amazon Mechanical Turk. As we saw in chapter 2, this distributed platform meant that it was suddenly possible to access a distributed labor force to do on-line tasks, like labeling and sorting images, at scale and at low cost. "He showed me the website, and I can tell you literally that day I knew the ImageNet project was going to happen," Li said. "Suddenly we found a tool that could scale, that we could not possibly dream of by hiring Princeton undergrads."[39] Un-surprisingly, the undergraduates did not get the job.

Instead, ImageNet would become, for a time, the world's largest academic user of Amazon's Mechanical Turk, deploy-ing an army of piecemeal workers to sort an average of fifty images a minute into thousands of categories.[40] There were categories for apples and airplanes, scuba divers and sumo wrestlers. But there were cruel, offensive, and racist labels, too: photographs of people were classified into categories like "alcoholic," "ape-man," "crazy," "hooker," and "slant eye." All of these terms were imported from WordNet's lexical database and given to crowdworkers to pair with images. Over the course of a decade, ImageNet grew into a colossus of object recognition for machine learning and a powerfully important benchmark for the field. The approach of mass data extraction without consent and labeling by underpaid crowdworkers would become standard practice, and hundreds of new training datasets would follow ImageNet's lead. As we will see in the next chapter, these practices—and the labeled data they generated—eventually came back to haunt the project.

The End of Consent

The early years of the twenty-first century marked a shift away from consent-driven data collection. In addition to dispensing with the need for staged photo shoots, those responsible for assembling datasets presumed that the contents of the internet were theirs for the taking, beyond the need for agreements, signed releases, and ethics reviews. Now even more troubling practices of extraction began to emerge. For example, at the Colorado Springs campus of the University of Colorado, a professor installed a camera on the main walkway of the campus and secretly captured photos of more than seventeen hundred students and faculty—all to train a facial recognition system of his own.[41] A similar project at Duke University harvested footage of more than two thousand students without their knowledge as they went between their classes and then published the results on the internet. The dataset, called DukeMTMC (for multitarget, multicamera facial recognition), was funded by the U.S. Army Research Office and the National Science Foundation.[42]

The DukeMTMC project was roundly criticized after an investigative project by artists and researchers Adam Harvey and Jules LaPlace showed that the Chinese government was using the images to train systems for the surveillance of ethnic minorities. This spurred an investigation by Duke's institutional review board, which determined that this was a "significant deviation" from acceptable practices. The dataset was removed from the internet.[43]

But what happened at the University of Colorado and Duke were by no means isolated cases. At Stanford University, researchers commandeered a webcam from a popular café in San Francisco to extract almost twelve thousand images of "everyday life of a busy downtown café" without anyone's consent.[44] Over and over, data extracted without permission or consent would be uploaded for machine learning researchers, who would then use it as an infrastructure for automated imaging systems.

Another example is Microsoft's landmark training dataset MS-Celeb, which scraped approximately ten million photos of a hundred thousand celebrities from the internet in 2016. At the time, it was the largest public facial recognition dataset in the world, and the people included were not just famous actors and politicians but also journalists, activists, policymakers, academics, and artists.[45] Ironically, several of the people who had been included in the set without consent are known for their work critiquing surveillance and facial recognition itself, including documentary filmmaker Laura Poitras; digital rights activist Jillian York; critic Evgeny Morozov; and the author of *Surveillance Capitalism*, Shoshana Zuboff.[46]

Even when datasets are scrubbed of personal information and released with great caution, people have been re-identified or highly sensitive details about them have been revealed. In 2013, for example, the New York City Taxi and

Limousine Commission released a dataset of 173 million individual cab rides, and it included pickup and drop-off times,
locations, fares, and tip amounts. The taxi drivers' medallion
numbers were anonymized, but this was quickly undone, en-
abling researchers to infer sensitive information like annual in-
comes and home addresses.[47] Once combined with public information from sources like celebrity blogs, some actors and
politicians were identified, and it was possible to deduce the
addresses of people who visited strip clubs.[48] But beyond individual harms, such datasets also generate "predictive privacy
harms" for whole groups or communities.[49] For instance, the
same New York City taxi dataset was used to suggest which taxi
drivers were devout Muslims by observing when they stopped
at prayer times.[50]

From any seemingly innocuous and anonymized dataset can come many unexpected and highly personal forms of
information, but this fact has not hampered the collection of
images and text. As success in machine learning has come to
rely on ever-larger datasets, more people are seeking to acquire
them. But why does the wider AI field accept this practice, despite the ethical, political, and epistemological problems and
potential harms? What beliefs, justifications, and economic incentives normalized this mass extraction and general equivalence of data?

Myths and Metaphors of Data

The oft-cited history of artificial intelligence written by AI
professor Nils Nilsson outlines several of the founding myths
about data in machine learning. He neatly illustrates how data
is typically described in the technical disciplines: "The great
volume of raw data calls for efficient 'data-mining' techniques
for classifying, quantifying, and extracting useful information.

Machine learning methods are playing an increasingly important role in data analysis because they can deal with massive amounts of data. In fact, the more data the better."[51]

Echoing Robert Mercer from decades earlier, Nilsson perceived that data was everywhere for the taking, and all the better for mass classification by machine learning algorithms.[52] It was such a common belief as to have become axiomatic: data is there to be acquired, refined, and made valuable.

But vested interests carefully manufactured and supported this belief over time. As sociologists Marion Fourcade and Kieran Healy note, the injunction always to collect data came not only from the data professions but also from their institutions and the technologies they deploy:

> The institutional command coming from technology is the most potent of all: we do these things *because we can.* . . . Professionals recommend, the institutional environment demands, and technology enables organizations to sweep up as much individual data as possible. It does not matter that the amounts collected may vastly exceed a firm's imaginative reach or analytic grasp. The assumption is that it will eventually be useful, i.e. valuable. . . . Contemporary organizations are both culturally impelled by the data imperative and powerfully equipped with new tools to enact it.[53]

This produced a kind of moral imperative to collect data in order to make systems better, regardless of the negative impacts the data collection might cause at any future point. Behind the questionable belief that "more is better" is the idea that individuals can be completely knowable, once enough disparate pieces of data are collected.[54] But what counts as data?

Historian Lisa Gitelman notes that every discipline and institution "has its own norms and standards for the imagination of data."[55] Data, in the twenty-first century, became whatever could be captured.

Terms like "data mining" and phrases like "data is the new oil" were part of a rhetorical move that shifted the notion of data away from something personal, intimate, or subject to individual ownership and control toward something more inert and nonhuman. Data began to be described as a resource to be consumed, a flow to be controlled, or an investment to be harnessed.[56] The expression "data as oil" became commonplace, and although it suggested a picture of data as a crude material for extraction, it was rarely used to emphasize the costs of the oil and mining industries: indentured labor, geopolitical conflicts, depletion of resources, and consequences stretching beyond human timescales.

Ultimately, "data" has become a bloodless word; it disguises both its material origins and its ends. And if data is seen as abstract and immaterial, then it more easily falls outside of traditional understandings and responsibilities of care, consent, or risk. As researchers Luke Stark and Anna Lauren Hoffman argue, metaphors of data as a "natural resource" just lying in wait to be discovered are a well-established rhetorical trick used for centuries by colonial powers.[57] Extraction is justified if it comes from a primitive and "unrefined" source.[58] If data is framed as oil, just waiting to be extracted, then machine learning has come to be seen as its necessary refinement process.

Data also started to be viewed as capital, in keeping with the broader neoliberal visions of markets as the primary forms of organizing value. Once human activities are expressed through digital traces and then tallied up and ranked within scoring metrics, they function as a way to extract value. As

Fourcade and Healy observe, those who have the right data signals gain advantages like discounted insurance and higher standing across markets.[59] High achievers in the mainstream economy tend to do well in a data-scoring economy, too, while those who are poorest become targets of the most harmful forms of data surveillance and extraction. When data is considered as a form of capital, then everything is justified if it means collecting more. The sociologist Jathan Sadowski similarly argues that data now operates as a form of capital. He suggests that once everything is understood as data, it justifies a cycle of ever-increasing data extraction: "Data collection is thus driven by the perpetual cycle of capital accumulation, which in turn drives capital to construct and rely upon a world in which everything is made of data. The supposed universality of data reframes everything as falling under the domain of data capitalism. All spaces must be subjected to datafication. If the universe is conceived of as a potentially infinite reserve of data, then that means the accumulation and circulation of data can be sustained forever."[60]

This drive to accumulate and circulate is the powerful underlying ideology of data. Mass data extraction is the "new frontier of accumulation and next step in capitalism," Sadowski suggests, and it is the foundational layer that makes AI function.[61] Thus, there are entire industries, institutions, and individuals who don't want this frontier—where data is there for the taking—to be questioned or destabilized.

Machine learning models require ongoing flows of data to become more accurate. But machines are asymptotic, never reaching full precision, which propels the justification for more extraction from as many people as possible to fuel the refineries of AI. This has created a shift away from ideas like "human subjects"—a concept that emerged from the ethics debates of the twentieth century—to the creation of "data subjects," agglomerations of data points without subjectivity or context or clearly defined rights.

Ethics at Arm's Length

The great majority of university-based AI research is done without any ethical review process. But if machine learning techniques are being used to inform decisions in sensitive domains like education and health care, then why are they not subject to greater review? To understand that, we need to look at the precursor disciplines of artificial intelligence. Before the emergence of machine learning and data science, the fields of applied mathematics, statistics, and computer science had not historically been considered forms of research on human subjects.

In the early decades of AI, research using human data was usually seen to be a minimal risk.[62] Even though datasets in machine learning often come from and represent people and their lives, the research that used those datasets was seen more as a form of applied math with few consequences for human subjects. The infrastructures of ethics protections, like university-based institutional review boards (IRBs), had accepted this position for years.[63] This initially made sense; IRBs had been overwhelmingly focused on the methods common to biomedical and psychological experimentation in which interventions carry clear risks to individual subjects. Computer science was seen as far more abstract.

Once AI moved out of the laboratory contexts of the 1980s and 1990s and into real-world situations—such as attempting to predict which criminals will reoffend or who should receive welfare benefits—the potential harms expanded. Further, those harms affect entire communities as well as individuals. But there is still a strong presumption that publicly available datasets pose minimal risks and therefore should be exempt from ethics review.[64] This idea is the product of an earlier era, when it was harder to move data between locations and very expensive to store it for long periods. Those earlier assumptions are out of step with what is currently going on in machine learning. Now datasets are more easily connectable, indefinitely repurposable, continuously updatable, and frequently removed from the context of collection.

The risk profile of AI is rapidly changing as its tools become more invasive and as researchers are increasingly able to access data without interacting with their subjects. For example, a group of machine learning researchers published a paper in which they claimed to have developed an "automatic system for classifying crimes."[65] In particular, their focus was on whether a violent crime was gang-related, which they claimed their neural network could predict with only four pieces of information: the weapon, the number of suspects, the neighborhood, and the location. They did this using a crime dataset from the Los Angeles Police Department, which included thousands of crimes that had been labeled by police as gang-related.

Gang data is notoriously skewed and riddled with errors, yet researchers use this database and others like it as a definitive source for training predictive AI systems. The CalGang database, for example, which is widely used by police in California, has been shown to have major inaccuracies. The state auditor discovered that 23 percent of the hundreds of records it reviewed lacked adequate support for inclusion. The database also contained forty-two infants, twenty-eight of whom were listed for having "admitting to being gang members."[66] Most of the adults on the list had never been charged, but once they were included in the database, there was no way to have their name removed. Reasons for being included might be as simple as chatting with a neighbor while wearing a red shirt; using these trifling justifications, Black and Latinx people have been disproportionately added to the list.[67]

When the researchers presented their gang-crime prediction project at a conference, some attendees were troubled. As reported by *Science,* questions from the audience included, "How could the team be sure the training data were not biased to begin with?" and "What happens when someone is mislabeled as a gang member?" Hau Chan, a computer scientist now at Harvard University who presented the work, responded that he couldn't know how the new tool would be used. "[These are the] sort of ethical questions that I don't know how to answer appropriately," he said, being just "a researcher." An audience member replied by quoting a lyric from Tom Lehrer's satiric song about the wartime rocket scientist Wernher von Braun: "Once the rockets are up, who cares where they come down?"[68]

This separation of ethical questions away from the technical reflects a wider problem in the field, where the responsibility for harm is either not recognized or seen as beyond the scope of the research. As Anna Lauren Hoffman writes: "The problem here isn't only one of biased datasets or unfair algorithms and of unintended consequences. It's also indicative of a more persistent problem of researchers actively reproducing ideas that damage vulnerable communities and reinforce current injustices. Even if the Harvard team's proposed system for identifying gang violence is never implemented, hasn't a kind of damage already been done? Wasn't their project an act of cultural violence in itself?"[69] Sidelining issues of ethics is harmful in itself, and it perpetuates the false idea that scientific research happens in a vacuum, with no responsibility for the ideas it propagates.

The reproduction of harmful ideas is particularly dangerous now that AI has moved from being an experimental discipline used only in laboratories to being tested at scale on millions of people. Technical approaches can move rapidly from conference papers to being deployed in production systems, where harmful assumptions can become ingrained and hard to reverse.

Machine learning and data-science methods can create an abstract relationship between researchers and subjects, where work is being done at a distance, removed from the communities and individuals at risk of harm. This arm's-length relationship of AI researchers to the people whose lives are reflected in datasets is a long-established practice. Back in 1976, when AI scientist Joseph Weizenbaum wrote his scathing critique of the field, he observed that computer science was already seeking to circumvent all human contexts.[70] He argued that data systems allowed scientists during wartime to operate at a psychological distance from the people "who would be maimed and killed by the weapons systems that would result from the ideas they communicated."[71] The answer, in Weizenbaum's view, was to directly contend with what data actually represents: "The lesson, therefore, is that the scientist and technologist must, by acts of will and of the imagination, actively strive to reduce such psychological distances, to counter the forces that tend to remove him from the consequences of his actions. He must—it is as simple as this—think of what he is actually doing."[72]

Weizenbaum hoped that scientists and technologists would think more deeply about the consequences of their work—and of who might be at risk. But this would not become the standard of the AI field. Instead, data is more commonly seen as something to be taken at will, used without restriction, and interpreted without context. There is a rapacious international culture of data harvesting that can be exploitative and invasive and can produce lasting forms of harm.[73] And there are many industries, institutions, and individuals who are strongly incentivized to maintain this colonizing attitude—where data is there for the taking—and they do not want it questioned or regulated.

The Capture of the Commons

The current widespread culture of data extraction continues to grow despite concerns about privacy, ethics, and safety. By researching the thousands of datasets that are freely available for AI development, I got a glimpse into what technical systems are built to recognize, of how the world is rendered for computers in ways that humans rarely see. There are gigantic datasets full of people's selfies, tattoos, parents walking with their children, hand gestures, people driving their cars, people committing crimes on CCTV, and hundreds of everyday human actions like sitting down, waving, raising a glass, or crying. Every form of biodata—including forensic, biometric, sociometric, and psychometric—is being captured and logged into databases for AI systems to find patterns and make assessments.

Training sets raise complex questions from ethical, methodological, and epistemological perspectives. Many were made without people's knowledge or consent and were harvested from online sources like Flickr, Google image search, and You-Tube or were donated by government agencies like the FBI. This data is now used to expand facial recognition systems, modulate health insurance rates, penalize distracted drivers, and fuel predictive policing tools. But the practices of data extraction are extending even deeper into areas of human life that were once off-limits or too expensive to reach. Tech companies have drawn on a range of approaches to gain new ground. Voice data is gathered from devices that sit on kitchen counters or bedroom nightstands; physical data comes from watches on wrists and phones in pockets; data about what books and newspapers are read comes from tablets and laptops; gestures and facial expressions are compiled and assessed in workplaces and classrooms.

The collection of people's data to build AI systems raises clear privacy concerns. Take, for example, the deal that Britain's Royal Free National Health Service Foundation Trust made with Google's subsidiary DeepMind to share the patient data records of 1.6 million people. The National Health Service in Britain is a revered institution, entrusted to provide health care that is primarily free to all while keeping patient data secure. But when the agreement with DeepMind was investigated, the company was found to have violated data protection laws by not sufficiently informing patients.[74] In her findings, the information commissioner observed that "the price of innovation does not need to be the erosion of fundamental privacy rights."[75]

Yet there are other serious issues that receive less attention than privacy. The practices of data extraction and training dataset construction are premised on a commercialized capture of what was previously part of the commons. This particular form of erosion is a privatization by stealth, an extraction of knowledge value from public goods. A dataset may still be publicly available, but the metavalue of the data—the model created by it—is privately held. Certainly, many good things can be done with public data. But there has been a social and, to some degree, a technical expectation that the value of data shared via public institutions and public spaces online should come back to the public good in other forms of the commons. Instead, we see a handful of privately owned companies that now have enormous power to extract insights and profits from those sources. The new AI gold rush consists of enclosing different fields of human knowing, feeling, and action—every type of available data—all caught in an expansionist logic of never-ending collection. It has become a pillaging of public space.

Fundamentally, the practices of data accumulation over many years have contributed to a powerful extractive logic, a logic that is now a core feature of how the AI field works. This logic has enriched the tech companies with the largest data pipelines, while the spaces free from data collection have dramatically diminished. As Vannevar Bush foresaw, machines have enormous appetites. But how and what they are fed has an enormous impact on how they will interpret the world, and the priorities of their masters will always shape how that vision is monetized. By looking at the layers of training data that shape and inform AI models and algorithms, we can see that gathering and labeling data about the world is a social and political intervention, even as it masquerades as a purely technical one.

The way data is understood, captured, classified, and named is fundamentally an act of world-making and containment. It has enormous ramifications for the way artificial intelligence works in the world and which communities are most affected. The myth of data collection as a benevolent practice in computer science has obscured its operations of power, protecting those who profit most while avoiding responsibility for its consequences.

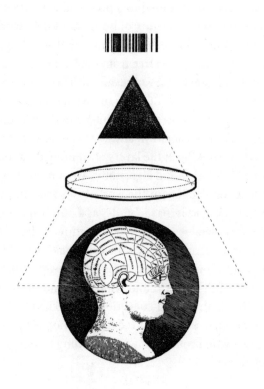

4

Classification

I am surrounded by human skulls. This room contains almost five hundred, collected in the early decades of the 1800s. All are varnished, with numbers inscribed in black ink on the frontal bone. Delicate calligraphic circles mark out areas of the skull associated in phrenology with particular qualities, including "Benevolence" and "Veneration." Some bear descriptions in capital letters, with words like "Dutchman," "Peruvian of the Inca Race," or "Lunatic." Each was painstakingly weighed, measured, and labeled by the American craniologist Samuel Morton. Morton was a physician, natural historian, and member of the Academy of Natural Sciences of Philadelphia. He gathered these human skulls from around the world by trading with a network of scientists and skull hunters who brought back specimens for his experiments, sometimes by robbing graves.[1] By the end of his life in 1851, Morton had amassed more than a thousand skulls, the largest collection in the world at the time.[2] Much of the archive is now held in storage at the Physical Anthropology Section of the Penn Museum in Philadelphia.

Morton was not a classical phrenologist in that he didn't believe that human character could be read through examin-

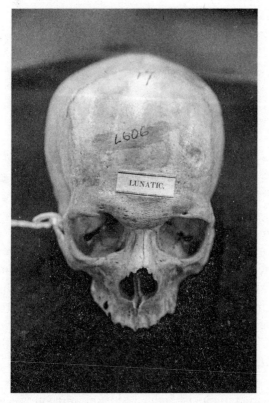

A skull from the Morton cranial collection marked "Lunatic."
Photograph by Kate Crawford

ing the shape of the head. Rather, his aim was to classify and
rank human races "objectively" by comparing the physical
characteristics of skulls. He did this by dividing them into the
five "races" of the world: African, Native American, Caucasian, Malay, and Mongolian—a typical taxonomy of the time
and a reflection of the colonialist mentality that dominated its
geopolitics.[3] This was the viewpoint of polygenism—the belief
that distinct human races had evolved separately at different

times—legitimized by white European and American scholars and hailed by colonial explorers as a justification for racist violence and dispossession.[4] Craniometry grew to be one of their leading methods since it purported to assess human difference and merit accurately.[5]

Many of the skulls I see belong to people who were born in Africa but who died enslaved in the Americas. Morton measured these skulls by filling the cranial cavities with lead shot, then pouring the shot back into cylinders and gauging the volume of lead in cubic inches.[6] He published his results, comparing them to skulls he acquired from other locations: for example, he claimed that white people had the largest skulls, while Black people were on the bottom of the scale. Morton's tables of average skull volume by race were regarded as the cutting edge of science of the time. His work was cited for the rest of the century as objective, hard data that proved the relative intelligence of human races and biological superiority of the Caucasian race. This research was instrumented in the United States to maintain the legitimacy of slavery and racial segregation.[7] Considered the scientific state of the art at the time, it was used to authorize racial oppression long after the studies were no longer cited.

But Morton's work was not the kind of evidence it claimed to be. As Stephen Jay Gould describes in his landmark book *The Mismeasure of Man*:

> In short, and to put it bluntly, Morton's summaries are a patchwork of fudging and finagling in the clear interest of controlling *a priori* convictions. Yet—and this is the most intriguing aspect of his case—I find no evidence of conscious fraud. . . . The prevalence of unconscious finagling, on the other hand, suggests a general conclusion about the so-

cial context of science. For if scientists can be hon-
estly self-deluded to Morton's extent, then prior
prejudice may be found anywhere, even in the ba-
sics of measuring bones and toting sums.[8]

Gould, and many others since, has reweighed the skulls
and reexamined Morton's evidence.[9] Morton made errors
and miscalculations, as well as procedural omissions, such as
ignoring the basic fact that larger people have larger brains.[10]
He selectively chose samples that supported his belief of white
supremacy and deleted the subsamples that threw off his group
averages. Contemporary assessments of the skulls at the Penn
Museum show no significant differences among people—even
when using Morton's data.[11] But prior prejudice—a way of seeing the world—had shaped what Morton believed was objective science and was a self-reinforcing loop that influenced his
findings as much as the lead-filled skulls themselves.

Craniometry was, as Gould notes, "the leading numerical
science of biological determinism during the nineteenth century" and was based on "egregious errors" in terms of the core
underlying assumptions: that brain size equated to intelligence,
that there are separate human races which are distinct biological species, and that those races could be placed in a hierarchy
according to their intellect and innate character.[12] Ultimately,
this kind of race science was debunked, but as Cornel West
has argued, its dominant metaphors, logics, and categories not
only supported white supremacy but also made specific political ideas about race possible while closing down others.[13]

Morton's legacy foreshadows epistemological problems
with measurement and classification in artificial intelligence.
Correlating cranial morphology with intelligence and claims to
legal rights acts as a technical alibi for colonialism and slavery.[14]
While there is a tendency to focus on the errors in skull mea-

surements and how to correct for them, the far greater error is in the underlying worldview that animated this methodology. The aim, then, should be not to call for more accurate or "fair" skull measurements to shore up racist models of intelligence but to condemn the approach altogether. The practices of classification that Morton used were *inherently* political, and his invalid assumptions about intelligence, race, and biology had far-ranging social and economic effects.

The politics of classification is a core practice in artificial intelligence. The practices of classification inform how machine intelligence is recognized and produced from university labs to the tech industry. As we saw in the previous chapter, artifacts in the world are turned into data through extraction, measurement, labeling, and ordering, and this becomes—intentionally or otherwise—a slippery ground truth for technical systems trained on that data. And when AI systems are shown to produce discriminatory results along the categories of race, class, gender, disability, or age, companies face considerable pressure to reform their tools or diversify their data. But the result is often a narrow response, usually an attempt to address technical errors and skewed data to make the AI system appear more fair. What is often missing is a more fundamental set of questions: How does classification function in machine learning? What is at stake when we classify? In what ways do classifications interact with the classified? And what unspoken social and political theories underlie and are supported by these classifications of the world?

In their landmark study of classification, Geoffrey Bowker and Susan Leigh Star write that "classifications are powerful technologies. Embedded in working infrastructures they become relatively invisible without losing any of their power."[15] Classification is an act of power, be it labeling images in AI training sets, tracking people with facial recognition, or pouring lead shot into skulls. But classifications can disappear, as Bowker and Star observe, "into infrastructure, into habit, into the taken for granted."[16] We can easily forget that the classifications that are casually chosen to shape a technical system can play a dynamic role in shaping the social and material world.

The tendency to focus on the issue of bias in artificial intelligence has drawn us away from assessing the core practices of classification in AI, along with their attendant politics. To see that in action, in this chapter we'll explore some of the training datasets of the twenty-first century and observe how their schemas of social ordering naturalize hierarchies and magnify inequalities. We will also look at the limits of the bias debates in AI, where mathematical parity is frequently proposed to produce "fairer systems" instead of contending with underlying social, political, and economic structures. In short, we will consider how artificial intelligence uses classification to encode power.

Systems of Circular Logic

A decade ago, the suggestion that there could be a problem of bias in artificial intelligence was unorthodox. But now examples of discriminatory AI systems are legion, from gender bias in Apple's creditworthiness algorithms to racism in the COMPAS criminal risk assessment software and to age bias in Facebook's ad targeting.[17] Image recognition tools miscategorize Black faces, chatbots adopt racist and misogynistic language, voice recognition software fails to recognize female-sounding voices, and social media platforms show more highly paid job advertisements to men than to women.[18] As scholars like Ruha Benjamin and Safiya Noble have shown, there are hundreds of examples throughout the tech ecosystem.[19] Many more have never been detected or publicly admitted.

The typical structure of an episode in the ongoing AI bias narrative begins with an investigative journalist or whistle-blower revealing how an AI system is producing discriminatory results. The story is widely shared, and the company in question promises to address the issue. Then either the system is superseded by something new, or technical interventions are made in the attempt to produce results with greater parity. Those results and technical fixes remain proprietary and secret, and the public is told to rest assured that the malady of bias has been "cured."[20] It is much rarer to have a public debate about *why* these forms of bias and discrimination frequently recur and whether more fundamental problems are at work than simply an inadequate underlying dataset or a poorly designed algorithm.

One of the more vivid examples of bias in action comes from an insider account at Amazon. In 2014, the company decided to experiment with automating the process of recommending and hiring workers. If automation had worked to drive profits in product recommendation and warehouse organization, it could, the logic went, make hiring more efficient. In the words of one engineer, "They literally wanted it to be an engine where I'm going to give you 100 resumes, it will spit out the top five, and we'll hire those."[21] The machine learning system was designed to rank people on a scale of one to five, mirroring Amazon's system of product ratings. To build the underlying model, Amazon's engineers used a dataset of ten years' worth of résumés from fellow employees and then trained a statistical model on fifty thousand terms that appeared in those résumés. Quickly, the system began to assign less importance to commonly used engineering terms, like programming languages, because everyone listed them in their job histories. Instead, the models began valuing more subtle cues that recurred on successful applications. A strong preference emerged for particular verbs. The examples the engineers mentioned were "executed" and "captured."[22]

Recruiters starting using the system as a supplement to their usual practices.[23] Soon enough, a serious problem emerged: the system wasn't recommending women. It was actively downgrading résumés from candidates who attended women's colleges, along with any résumés that even included the word "women." Even after editing the system to remove the influence of explicit references to gender, the biases remained. Proxies for hegemonic masculinity continued to emerge in the gendered use of language itself. The model was biased against women not just as a category but against commonly gendered forms of speech.

Inadvertently, Amazon had created a diagnostic tool. The vast majority of engineers hired by Amazon over ten years had been men, so the models they created, which were trained on the successful résumés of men, had learned to recommend men for future hiring. The employment practices of the past and present were shaping the hiring tools for the future. Amazon's system unexpectedly revealed the ways bias already existed, from the way masculinity is encoded in language, in résumés, and in the company itself. The tool was an intensification of the existing dynamics of Amazon and highlighted the lack of diversity across the AI industry past and present.[24]

Amazon ultimately shut down its hiring experiment. But the scale of the bias problem goes much deeper than a single system or failed approach. The AI industry has traditionally understood the problem of bias as though it is a bug to be fixed rather than a feature of classification itself. The result has been a focus on adjusting technical systems to produce greater quantitative parity across disparate groups, which, as we'll see, has created its own problems.

Understanding the relation between bias and classification requires going beyond an analysis of the production of knowledge—such as determining whether a dataset is biased or unbiased—and, instead, looking at the mechanics of knowledge construction itself, what sociologist Karin Knorr Cetina calls the "epistemic machinery."[25] To see that requires observing how patterns of inequality across history shape access to resources and opportunities, which in turn shape data. That data is then extracted for use in technical systems for classification and pattern recognition, which produces results that are perceived to be somehow objective. The result is a statistical ouroboros: a self-reinforcing discrimination machine that amplifies social inequalities under the guise of technical neutrality.

The Limits of Debiasing Systems

To better understand the limitations of analyzing AI bias, we can look to the attempts to fix it. In 2019, IBM tried to respond to concerns about bias in its AI systems by creating what the company described as a more "inclusive" dataset called Diversity in Faces (DiF).[26] DiF was part of an industry response to the groundbreaking work released a year earlier by researchers Joy Buolamwini and Timnit Gebru that had demonstrated that several facial recognition systems—including those by IBM, Microsoft, and Amazon—had far greater error rates for people with darker skin, particularly women.[27] As a result, efforts were ongoing inside all three companies to show progress on rectifying the problem.

"We expect face recognition to work accurately for each of us," the IBM researchers wrote, but the only way that the "challenge of diversity could be solved" would be to build "a data set comprised from the face of every person in the world."[28] IBM's researchers decided to draw on a preexisting dataset of a hundred million images taken from Flickr, the largest publicly available collection on the internet at the time.[29] They then used one million photos as a small sample and measured the craniofacial distances between landmarks in each face: eyes, nasal width, lip height, brow height, and so on. Like Morton measuring skulls, the IBM researchers sought to assign cranial measures and create categories of difference.

The IBM team claimed that their goal was to increase diversity of facial recognition data. Though well intentioned, the classifications they used reveal the politics of what diversity meant in this context. For example, to label the gender and age of a face, the team tasked crowdworkers to make subjective annotations, using the restrictive model of binary gender. Anyone who seemed to fall outside of this binary was removed from the dataset. IBM's vision of diversity emphasized the expansive options for cranial orbit height and nose bridges but discounted the existence of trans or gender nonbinary people. "Fairness" was reduced to meaning higher accuracy rates for machine-led facial recognition, and "diversity" referred to a wider range of faces to train the model. Craniometric analysis functions like a bait and switch, ultimately depoliticizing the idea of diversity and replacing it with a focus on *variation*. Designers get to decide what the variables are and how people are allocated to categories. Again, the practice of classification is centralizing power: the power to decide which differences make a difference.

IBM's researchers go on to state an even more problematic conclusion: "Aspects of our heritage—including race, ethnicity, culture, geography—and our individual identity—age, gender and visible forms of self-expression—are reflected in our faces."[30] This claim goes against decades of research that has challenged the idea that race, gender, and identity are biological categories at all but are better understood as politically, culturally, and socially constructed.[31] Embedding identity claims in technical systems as though they are facts observable from the face is an example of what Simone Browne calls "digital epidermalization," the imposition of race on the body. Browne defines this as the exercise of power when the disembodied gaze of surveillance technologies "do the work of alienating the subject by producing a 'truth' about the body and one's identity (or identities) despite the subject's claims."[32]

The foundational problems with IBM's approach to classifying diversity grow out of this kind of centralized production of identity, led by the machine learning techniques that were available to the team. Skin color detection is done because it can be, not because it says anything about race or produces a deeper cultural understanding. Similarly, the use of cranial measurement is done because it is a method that *can* be done with machine learning. The affordances of the tools become the horizon of truth. The capacity to deploy cranial measurements and digital epidermalization at scale drives a desire to find meaning in these approaches, even if this method has nothing to do with culture, heritage, or diversity. They are used to increase a problematic understanding of accuracy. Technical claims about accuracy and performance are commonly shot through with political choices about categories and norms but are rarely acknowledged as such.[33] These approaches are grounded in an ideological premise of biology as destiny, where our faces become our fate.

The Many Definitions of Bias

Since antiquity, the act of classification has been aligned with power. In theology, the ability to name and divide things was a divine act of God. The word "category" comes from the Ancient Greek *katēgoría*, formed from two roots: *kata* (against) and *agoreuo* (speaking in public). In Greek, the word can be either a logical assertion or an accusation in a trial—alluding to both scientific and legal methods of categorization.

The historical lineage of "bias" as a term is much more recent. It first appears in fourteenth-century geometry, where it refers to an oblique or diagonal line. By the sixteenth century, it had acquired something like its current popular meaning, of "undue prejudice." By the 1900s, "bias" had developed a more technical meaning in statistics, where it refers to systematic differences between a sample and population, when the sample is not truly reflective of the whole.[34] It is from this statistical tradition that the machine learning field draws its understanding of bias, where it relates to a set of other concepts: generalization, classification, and variance.

Machine learning systems are designed to be able to generalize from a large training set of examples and to correctly classify new observations not included in the training datasets.[35] In other words, machine learning systems can perform a type of induction, learning from specific examples (such as past résumés of job applicants) in order to decide which data points to look for in new examples (such as word groupings in résumés from new applicants). In such cases, the term "bias" refers to a type of error that can occur during this predictive process of generalization—namely, a systematic or consistently reproduced classification error that the system exhibits when presented with new examples. This type of bias is often contrasted with another type of generalization error, variance, which refers to an algorithm's sensitivity to differences in training data. A model with high bias and low variance may be underfitting the data—failing to capture all of its significant features or signals. Alternatively, a model with high variance and low bias may be overfitting the data—building a model too close to the training data so that it potentially captures "noise" in addition to the data's significant features.[36]

Outside of machine learning, "bias" has many other meanings. For instance, in law, bias refers to a preconceived notion or opinion, a judgment based on prejudices, as opposed to a decision come to from the impartial evaluation of the facts of a case.[37] In psychology, Amos Tversky and Daniel Kahneman study "cognitive biases," or the ways in which human judgments deviate systematically from probabilistic expectations.[38] More recent research on implicit biases emphasizes the ways that unconscious attitudes and stereotypes "produce behaviors that diverge from a person's avowed or endorsed beliefs or principles."[39] Here bias is not simply a type of technical error; it also opens onto human beliefs, stereotypes, or forms of discrimination. These definitional distinctions limit the utility of "bias" as a term, especially when used by practitioners from different disciplines.

Technical designs can certainly be improved to better account for how their systems produce skews and discriminatory results. But the harder questions of why AI systems perpetuate forms of inequity are commonly skipped over in the rush to arrive at narrow technical solutions of statistical bias as though that is a sufficient remedy for deeper structural problems. There has been a general failure to address the ways in which the instruments of knowledge in AI reflect and serve the incentives of a wider extractive economy. What remains is a persistent asymmetry of power, where technical systems maintain and extend structural inequality, regardless of the intention of the designers.

Every dataset used to train machine learning systems, whether in the context of supervised or unsupervised machine learning, whether seen to be technically biased or not, contains a worldview. To create a training set is to take an almost infinitely complex and varied world and fix it into taxonomies composed of discrete classifications of individual data points, a process that requires inherently political, cultural, and social choices. By paying attention to these classifications, we can glimpse the various forms of power that are built into the architectures of AI world-building.

Training Sets as Classification Engines: The Case of ImageNet

In the last chapter we looked at the history of ImageNet and how this benchmark training set has influenced computer vision research since its creation in 2009. By taking a closer look at ImageNet's structure, we can begin to see how the dataset is ordered and its underlying logic for mapping the world of objects. ImageNet's structure is labyrinthine, vast, and filled with curiosities. The underlying semantic structure of ImageNet was imported from WordNet, a database of word classifications first developed at Princeton University's Cognitive Science Laboratory in 1985 and funded by the U.S. Office of Naval Research.[40] WordNet was conceived as a machine-readable dictionary, where users would search on the basis of semantic rather than alphabetic similarity. It became a vital source for the fields of computational linguistics and natural language processing. The WordNet team collected as many words as they could, starting with the Brown Corpus, a collection of one million words compiled in the 1960s.[41] The words in the Brown Corpus came from newspapers and a ramshackle collection of books including *New Methods of Parapsychology, The Family Fallout Shelter,* and *Who Rules the Marriage Bed?*[42]

WordNet attempts to organize the entire English language into synonym sets, or synsets. The ImageNet researchers selected only nouns, with the idea that nouns are things that pictures can represent—and that would be sufficient to train machines to automatically recognize objects. So Image-

Net's taxonomy is organized according to a nested hierarchy derived from WordNet, in which each synset represents a distinct concept, with synonyms grouped together (for example, "auto" and "car" are treated as belonging to the same set). The hierarchy moves from more general concepts to more specific ones. For example, the concept "chair" is found under artifact → furnishing → furniture → seat → chair. This classification system unsurprisingly evokes many prior taxonomical ranks, from the Linnaean system of biological classification to the ordering of books in libraries.

But the first indication of the true strangeness of Image-Net's worldview is its nine top-level categories that it drew from WordNet: plant, geological formation, natural object, sport, artifact, fungus, person, animal, and miscellaneous. These are curious categories into which all else must be ordered. Below that, it spawns into thousands of strange and specific nested classes, into which millions of images are housed like Russian dolls. There are categories for apples, apple butter, apple dumplings, apple geraniums, apple jelly, apple juice, apple maggots, apple rust, apple trees, apple turnovers, apple carts, and apple-sauce. There are pictures of hot lines, hot pants, hot plates, hot pots, hot rods, hot sauce, hot springs, hot toddies, hot tubs, hot-air balloons, hot fudge sauce, and hot water bottles. It is a riot of words, ordered into strange categories like those from Jorge Luis Borges's mythical encyclopedia.[43] At the level of images, it looks like madness. Some images are high-resolution stock photography, others are blurry phone photographs in poor lighting. Some are photos of children. Others are stills from pornography. Some are cartoons. There are pin-ups, religious icons, famous politicians, Hollywood celebrities, and Italian comedians. It veers wildly from the professional to the amateur, the sacred to the profane.

Human classifications are a good place to see these politics of classification at work. In ImageNet the category "human body" falls under the branch Natural Object → Body → Human Body. Its subcategories include "male body," "person," "juvenile body," "adult body," and "female body." The "adult body" category contains the subclasses "adult female body" and "adult male body." There is an implicit assumption here that only "male" and "female" bodies are recognized as "natural." There is an ImageNet category for the term "Hermaphrodite," but it is situated within the branch Person → Sensualist → Bisexual alongside the categories "Pseudohermaphrodite" and "Switch Hitter."[44]

Even before we look at the more controversial categories within ImageNet, we can see the politics of this classificatory scheme. The decisions to classify gender in this way are also naturalizing gender as a biological construct, which is binary, and transgender or gender nonbinary people are either nonexistent or placed under categories of sexuality.[45] Of course, this is not a novel approach. The classification hierarchy of gender and sexuality in ImageNet recalls earlier harmful forms of categorization, such as the classification of homosexuality as a mental disorder in the *Diagnostic and Statistical Manual*.[46] This deeply damaging categorization was used to justify subjecting people to repressive so-called therapies, and it took years of activism before the American Psychiatric Association removed it in 1973.[47]

Reducing humans into binary gender categories and rendering transgender people invisible or "deviant" are common features of classification schemes in machine learning. Os Keyes's study of automatic gender detection in facial recognition shows that almost 95 percent of papers in the field treat gender as binary, with the majority describing gender as immutable and physiological.[48] While some might respond that this can be easily remedied by creating more categories, this fails to address the deeper harm of allocating people into gender or race categories without their input or consent. This practice has a long history. Administrative systems for centuries have sought to make humans legible by applying fixed labels and definite properties. The work of essentializing and ordering on the basis of biology or culture has long been used to justify forms of violence and oppression.

While these classifying logics are treated as though they are natural and fixed, they are moving targets: not only do they affect the people being classified, but how they impact people in turn changes the classifications themselves. Hacking calls this the "looping effect," produced when the sciences engage in "making up people."[49] Bowker and Star also underscore that once classifications of people are constructed, they can stabilize a contested political category in ways that are difficult to see.[50] They become taken for granted unless they are actively resisted. We see this phenomenon in the AI field when highly influential infrastructures and training datasets pass as purely technical, whereas in fact they contain political interventions within their taxonomies: they naturalize a particular ordering of the world which produces effects that are seen to justify their original ordering.

The Power to Define "Person"

To impose order onto an undifferentiated mass, to ascribe phenomena to a category—that is, to name a thing—is in turn a means of reifying the existence of that category.

In the case of the 21,841 categories that were originally in the ImageNet hierarchy, noun classes such as "apple" or "apple butter" might seem reasonably uncontroversial, but not all nouns are created equal. To borrow an idea from linguist George Lakoff, the concept of an "apple" is a more *nouny* noun than the concept of "light," which in turn is more nouny than a concept such as "health."[51] Nouns occupy various places on an axis from the concrete to the abstract, from the descriptive to the judgmental. These gradients have been erased in the logic of ImageNet. Everything is flattened out and pinned to a label, like taxidermy butterflies in a display case. While this approach has the aesthetics of objectivity, it is nonetheless a profoundly ideological exercise.

For a decade, ImageNet contained 2,832 subcategories under the top-level category "Person." The subcategory with the most associated pictures was "gal" (with 1,664 images) followed by "grandfather" (1,662), "dad" (1,643), and chief executive officer (1,614—most of them male). With these highly populated categories, we can already begin to see the outlines of a worldview. ImageNet contains a profusion of classificatory categories, including ones for race, age, nationality, profession, economic status, behavior, character, and even morality.

There are many problems with the way ImageNet's taxonomy purports to classify photos of people with the logics of object recognition. Even though its creators removed some explicitly offensive synsets in 2009, categories remained for racial and national identities including Alaska Native, Anglo-American, Black, Black African, Black Woman (but not White Woman), Latin American, Mexican American, Nicaraguan, Pakistani, South American Indian, Spanish American, Texan, Uzbek, White, and Zulu. To present these as logical categories of organizing people is already troubling, even before they are used to classify people based on their appearance. Other people are labeled by careers or hobbies: there are Boy Scouts, cheerleaders, cognitive neuroscientists, hairdressers, intelligence analysts, mythologists, retailers, retirees, and so on. The existence of these categories suggests that people can be visually ordered according to their profession, in a way that seems reminiscent of such children's books as Richard Scarry's *What Do People Do All Day?* ImageNet also contains categories that make no sense whatsoever for image classification such as Debtor, Boss, Acquaintance, Brother, and Color-Blind Person. These are all nonvisual concepts that describe a relationship, be it to other people, to a financial system, or to the visual field itself. The dataset reifies these categories and connects them to images, so that similar images can be "recognized" by future systems.

Many truly offensive and harmful categories hid in the depths of ImageNet's Person categories. Some classifications were misogynist, racist, ageist, and ableist. The list includes Bad Person, Call Girl, Closet Queen, Codger, Convict, Crazy, Deadeye, Drug Addict, Failure, Flop, Fucker, Hypocrite, Jezebel, Kleptomaniac, Loser, Melancholic, Nonperson, Pervert, Prima Donna, Schizophrenic, Second-Rater, Slut, Spastic, Spinster, Streetwalker, Stud, Tosser, Unskilled Person, Wanton, Waverer, and Wimp. Insults, racist slurs, and moral judgments abound.

These offensive terms remained in ImageNet for ten years. Because ImageNet was typically used for object recognition — with "object" broadly defined — the specific Person category was rarely discussed at technical conferences, nor did it receive much public attention until the ImageNet Roulette project went viral in 2019: led by the artist Trevor Paglen, the project included an app that allowed people to upload images to see how they would be classified based on ImageNet's Person categories.[52] This focused considerable media attention on the influential collection's longtime inclusion of racist and sexist terms. The creators of ImageNet published a paper shortly afterward titled "Toward Fairer Datasets" that sought to "remove unsafe synsets." They asked twelve graduate students to flag any categories that seemed unsafe because they were either "inherently offensive" (for example, containing profanity or "racial or gender slurs") or "sensitive" (not inherently offensive but terms that "may cause offense when applied inappropriately, such as the classification of people based on sexual orientation and religion").[53] While this project sought to assess the offensiveness of ImageNet's categories by asking graduate students, the authors nonetheless continue to support the automated classification of people based on photographs despite the notable problems.

The ImageNet team ultimately removed 1,593 of 2,832 of the People categories—roughly 56 percent—deeming them "unsafe," along with the associated 600,040 images. The remaining half-million images were "temporarily deemed safe."[54] But what constitutes *safe* when it comes to classifying people? The focus on the hateful categories is not wrong, but it avoids addressing questions about the workings of the larger system. The entire taxonomy of ImageNet reveals the complexities and dangers of human classification. While terms like "microeconomist" or "basketball player" may initially seem less concerning than the use of labels like "spastic," "unskilled person," "mulatto," or "redneck," when we look at the people who are labeled in these categories we see many assumptions and stereotypes, including race, gender, age, and ability. In the metaphysics of ImageNet, there are separate image categories for "assistant professor" and "associate professor"—as though once someone gets a promotion, her or his biometric profile would reflect the change in rank.

In fact, there are no neutral categories in ImageNet, because the selection of images always interacts with the meaning of words. The politics are baked into the classificatory logic, even when the words aren't offensive. ImageNet is a lesson, in this sense, of what happens when people are categorized like objects. But this practice has only become more common in recent years, often inside the tech companies. The classification schemes used in companies like Facebook are much harder to investigate and criticize: proprietary systems offer few ways for outsiders to probe or audit how images are ordered or interpreted.

Then there is the issue of where the images in ImageNet's Person categories come from. As we saw in the last chapter, ImageNet's creators harvested images en masse from image search engines like Google, extracted people's selfies and vacation photos without their knowledge, and then paid Mechanical Turk workers to label and repackage them. All the skews and biases in how search engines return results are then informing the subsequent technical systems that scrape and label them. Low-paid crowdworkers are given the impossible task of making sense of the images at the rate of fifty per minute and fitting them into categories based on WordNet sysnets and Wikipedia definitions.[55] Perhaps it is no surprise that when we investigate the bedrock layer of these labeled images, we find that they are beset with stereotypes, errors, and absurdities. A woman lying on a beach towel is a "kleptomaniac," a teenager in a sports jersey is labeled a "loser," and an image of the actor Sigourney Weaver appears, classified as a "hermaphrodite."

Images—like all forms of data—are laden with all sorts of potential meanings, irresolvable questions, and contradictions. In trying to resolve these ambiguities, ImageNet's labels compress and simplify complexity. The focus on making training sets "fairer" by deleting offensive terms fails to contend with the power dynamics of classification and precludes a more thorough assessment of the underlying logics. Even if the worst examples are fixed, the approach is still fundamentally built on an extractive relationship with data that is divorced from the people and places from whence it came. Then it is rendered through a technical worldview that seeks to fuse together a form of singular objectivity from what are complex and varied cultural materials. The worldview of ImageNet is not unusual in this sense. In fact, it is typical of many AI training datasets, and it reveals many of the problems of top-down schemes that flatten complex social, cultural, political, and historical relations into quantifiable entities. This phenomenon is perhaps most obvious and insidious when it comes to the widespread efforts to classify people by race and gender in technical systems.

Constructing Race and Gender

By focusing on classification in AI, we can trace the ways that gender, race, and sexuality are falsely assumed to be natural, fixed, and detectable biological categories. Surveillance scholar Simone Browne observes, "There is a certain assumption with these technologies that categories of gender identity and race are clear cut, that a machine can be programmed to assign gender categories or determine what bodies and body parts should signify."[56] Indeed, the idea that race and gender can be automatically detectable in machine learning is treated as an assumed fact and rarely questioned by the technical disciplines, despite the profound political problems this presents.[57]

The UTKFace dataset (produced by a group at the University of Tennessee at Knoxville), for example, consists of more than twenty thousand images of faces with annotations for age, gender, and race.[58] The dataset's authors state that the dataset can be used for a variety of tasks, including automated face detection, age estimation, and age progression. The annotations for each image include an estimated age for each person, expressed in years from zero to 116. Gender is a forced binary: either zero for male or one for female. Second, race is categorized into five classes: White, Black, Asian, Indian, and Others. The politics of gender and race here are as obvious as they are harmful. Yet these kinds of dangerously reductive categorizations are widely used across many human-classifying training sets and have been part of the AI production pipelines for years.

UTKFace's narrow classificatory schema echoes the problematic racial classifications of the twentieth century, such as South Africa's apartheid system. As Bowker and Star have detailed, the South African government passed legislation in the 1950s that created a crude racial classification scheme to divide citizens into the categories of "Europeans, Asiatics, persons of mixed race or coloureds, and 'natives' or pure-blooded individuals of the Bantu race."[59] This racist legal regime governed people's lives, overwhelmingly those of Black South Africans whose movements were restricted and who were forcibly removed from their land. The politics of racial classification extended into the most intimate parts of people's lives. Interracial sexuality was forbidden, leading to more than 11,500 convictions by 1980, mostly of nonwhite women.[60] The complex centralized database for these classifications was designed and maintained by IBM, but the firm often had to rearrange the system and reclassify people, because in practice there were no singular pure racial categories.[61]

Above all, these systems of classification have caused enormous harm to people, and the concept of a pure "race" signifier has always been in dispute. In her writing about race, Donna Haraway observes, "In these taxonomies, which are, after all, little machines for clarifying and separating categories, the entity that always eluded the classifier was simple: race itself. The pure Type, which animated dreams, sciences, and terrors, kept slipping through, and endlessly multiplying, all the typological taxonomies."[62] Yet in dataset taxonomies, and in the machine learning systems that train on them, the myth of the pure type has emerged once more, claiming the authority of science. In an article on the dangers of facial

recognition, media scholar Luke Stark notes that "by intro-
ducing a variety of classifying logics that either reify existing
racial categories or produce new ones, the automated pattern-
generating logics of facial recognition systems both reproduce
systemic inequality and exacerbate it."[63]

Some machine learning methods go beyond predicting
age, gender, and race. There have been highly publicized efforts
to detect sexuality from photographs on dating sites and criminality based on headshots from drivers' licenses.[64] These approaches are deeply problematic for many reasons, not least of
which is that characteristics such as "criminality"—like race
and gender—are profoundly relational, socially determined
categories. These are not inherent features that are fixed; they
are contextual and shifting depending on time and place. To
make such predictions, machine learning systems are seeking to classify entirely relational things into fixed categories
and are rightly critiqued as scientifically and ethically problematic.[65]

Machine learning systems are, in a very real way, *constructing* race and gender: they are defining the world within
the terms they have set, and this has long-lasting ramifications
for the people who are classified. When such systems are hailed
as scientific innovations for predicting identities and future
actions, this erases the technical frailties of how the systems
were built, the priorities of why they were designed, and the
many political processes of categorization that shape them.
Disability scholars have long pointed to the ways in which so-
called normal bodies are classified and how that has worked
to stigmatize difference.[66] As one report notes, the history of
disability itself is a "story of the ways in which various systems
of classification (i.e., medical, scientific, legal) interface with
social institutions and their articulations of power and knowledge."[67] At multiple levels, the act of defining categories and

ideas of normalcy creates an outside: forms of abnormality, difference, and otherness. Technical systems are making political and normative interventions when they give names to something as dynamic and relational as personal identity, and they commonly do so using a reductive set of possibilities of what it is to be human. That restricts the range of how people are understood and can represent themselves, and it narrows the horizon of recognizable identities.

As Ian Hacking observes, classifying people is an imperial imperative: subjects were classified by empires when they were conquered, and then they were ordered into "a kind of people" by institutions and experts.[68] These acts of naming were assertions of power and colonial control, and the negative effects of those classifications can outlast the empires themselves. Classifications are technologies that produce and limit ways of knowing, and they are built into the logics of AI.

The Limits of Measurement

So what is to be done? If so much of the classificatory strata in training data and technical systems are forms of power and politics represented as objective measurement, how should we go about redressing this? How should system designers account for, in some cases, slavery, oppression, and hundreds of years of discrimination against some groups to the benefit of others? In other words, how should AI systems make representations of the social?

Making these choices about which information feeds AI systems to produce new classifications is a powerful moment of decision making: but who gets to choose and on what basis? The problem for computer science is that justice in AI systems will never be something that can be coded or computed. It requires a shift to assessing systems beyond optimization metrics and statistical parity and an understanding of where the frameworks of mathematics and engineering are causing the problems. This also means understanding how AI systems interact with data, workers, the environment, and the individuals whose lives will be affected by its use and deciding where AI should not be used.

Bowker and Star conclude that the sheer density of the collisions of classification schemes calls for a new kind of approach, a sensitivity to the "topography of things such as the distribution of ambiguity; the fluid dynamics of how classification systems meet up—a plate tectonics rather than static geology."[69] But it also requires attending to the uneven allocations of advantage and suffering, for "how these choices are made, and how we may think about that invisible matching process, is at the core of the ethical project."[70] Nonconsensual classifications present serious risks, as do normative assumptions about identity, yet these practices have become standard. That must change.

In this chapter we've seen how classificatory infrastructures contain gaps and contradictions: they necessarily reduce complexity, and they remove significant context, in order to make the world more computable. But they also proliferate in machine learning platforms in what Umberto Eco called "chaotic enumeration."[71] At a certain level of granularity, like and unlike things become sufficiently commensurate so that their similarities and differences are machine readable, yet in actuality their characteristics are uncontainable. Here, the issues go far beyond whether something is classified wrong or classified right. We are seeing strange, unpredictable twists as machine categories and people interact and change each other, as they try to find legibility in the shifting terrain, to fit the right categories and be spiked into the most lucrative feeds. In a machine learning landscape, these questions are no less urgent because they are hard to see. What is at stake is not just a historical curiosity or the odd feeling of a mismatch between the dotted-outline profiles we may glimpse in our platforms and feeds. Each and every classification has its consequence.

The histories of classification show us that the most harmful forms of human categorization—from the Apartheid system to the pathologization of homosexuality—did not simply fade away under the light of scientific research and ethical critique. Rather, change also required political organizing, sustained protest, and public campaigning over many years. Classificatory schemas enact and support the structures of power that formed them, and these do not shift without considerable effort. In Frederick Douglass's words, "Power concedes nothing without a demand. It never did and it never will."[72] Within the invisible regimes of classification in machine learning, it is harder to make demands and oppose their internal logics.

The training sets that are made public—such as ImageNet, UTKFace, and DiF—give us some insight into the kinds of categorizations that are propagating across industrial AI systems and research practices. But the truly massive engines of classification are the ones being operated at a global scale by private technology companies, including Facebook, Google, TikTok, and Baidu. These companies operate with little oversight into how they categorize and target users, and they fail to offer meaningful avenues for public contestation. When the matching processes of AI are truly hidden and people are kept unaware of why or how they receive forms of advantage or disadvantage, a collective political response is needed—even as it becomes more difficult.

5
Affect

In a remote outpost in the mountainous highlands of Papua New Guinea, a young American psychologist named Paul Ekman arrived with a collection of flash-cards and a new theory.[1] It was 1967, and Ekman had heard that the Fore people of Okapa were so isolated from the wider world that they would be his ideal test subjects. Like many Western researchers before him, Ekman had come to Papua New Guinea to extract data from the indigenous community. He was gathering evidence to bolster a controversial hypothesis: that all humans exhibit a small number of universal emotions or affects that are natural, innate, cross-cultural, and the same all over the world. Although that claim remains tenuous, it has had far-reaching consequences: Ekman's presuppositions about emotions have grown into an expanding industry worth well over seventeen billion dollars.[2] This is the story of how affect recognition came to be part of artificial intelligence and the problems this presents.

In the tropics of Okapa, guided by medical researcher D. Carleton Gajdusek and anthropologist E. Richard Sorenson, Ekman hoped to run experiments that would assess how the Fore recognized emotions conveyed by facial expressions. Because the Fore had minimal contact with Westerners or mass media, Ekman theorized that their recognition and display of core expressions would prove that such expressions were universal. His methods were simple. He would show them flashcards of facial expressions and see if they described the emotion as he did. In Ekman's own words, "All I was doing was showing funny pictures."[3]

But Ekman had no training in Fore history, language, culture, or politics. His attempts to conduct his flashcard experiments using translators floundered; he and his subjects were exhausted by the process, which he described as like pulling teeth.[4] Ekman left Papua New Guinea, frustrated by his first attempt at cross-cultural research on emotional expression. But this would just be the beginning.

Today affect recognition tools can be found in national security systems and at airports, in education and hiring startups, from systems that purport to detect psychiatric illness to policing programs that claim to predict violence. By looking at the history of how computer-based emotion detection came to be, we can understand how its methods have raised both ethical concerns and scientific doubts. As we will see, the claim that a person's interior state of feeling can be accurately assessed by analyzing their face is premised on shaky evidence.[5] In fact, a comprehensive review of the available scientific literature on inferring emotions from facial movements published in 2019 was definitive: there is *no reliable evidence* that you can accurately predict someone's emotional state from their face.[6]

How did this collection of contested claims and experimental methodologies resolve into an approach that drives many parts of the affect AI industry? Why did the idea that there is a small set of universal emotions, readily interpreted from the face, become so accepted in the AI field, despite considerable evidence to the contrary? To understand that requires tracing how these ideas developed, long before AI emotion detection tools were built into the infrastructure of everyday life.

Ekman is just one of many people who have contributed to the theories behind affect recognition. But the rich and surprising history of Ekman's research illuminates some of the complex forces driving the field. His work is connected to U.S. intelligence funding of the human sciences during the Cold War through foundational work in the field of computer vision to the post-9/11 security programs employed to identify terrorists and right up to the current fashion for AI-based emotion recognition. It is a chronicle that combines ideology, economic policy, fear-based politics, and the desire to extract more information about people than they are willing to give.

Emotion Prophets: When Feelings Pay

For the world's militaries, corporations, intelligence agencies, and police forces, the idea of automated affect recognition is as compelling as it is lucrative. It holds the promise of reliably filtering friend from foe, distinguishing lies from truths, and using the instruments of science to see into interior worlds.

Technology companies have captured immense volumes of surface-level imagery of human expressions—including billions of Instagram selfies, Pinterest portraits, TikTok videos, and Flickr photos. One of the many things made possible by this profusion of images is the attempt to extract the so-called hidden truth of interior emotional states using machine learning. Affect recognition is being built into several facial recognition platforms, from the biggest tech companies to small startups. Whereas facial recognition attempts to identify a *particular* individual, affect detection aims to detect and classify emotions by analyzing *any* face. These systems may not be doing what they purport to do, but they can nonetheless be powerful agents in influencing behavior and training people to perform in recognizable ways. These systems are already playing a role in shaping how people behave and how social institutions operate, despite a lack of substantial scientific evidence that they work.

Automated affect detection systems are now widely deployed, particularly in hiring. A startup in London called Human uses emotion recognition to analyze video interviews of job candidates. According to a report in the *Financial Times,* "The company claims it can spot the emotional expressions of prospective candidates and match them with personality traits"; the company then scores subjects on such personality traits as honesty or passion for a job.[7] The AI hiring company HireVue, which lists among its clients Goldman Sachs, Intel and Unilever, uses machine learning to assess facial cues to infer people's suitability for a job. In 2014, the company launched its AI system to extract microexpressions, tone of voice, and other variables from video job interviews, which they used to compare job applicants against the company's top performers.[8]

In January 2016, Apple acquired the startup Emotient, which claimed to have produced software capable of detecting emotions from images of faces.[9] Emotient grew out of academic research conducted at the University of California San Diego and is one of a number of startups working in this area.[10] Perhaps the largest of these is Affectiva, a company based in Boston that emerged from academic work done at Massachusetts Institute of Technology. At MIT, Rosalind Picard and her colleagues were part of an emergent wider field known as affective computing, which describes computing that "relates to, arises from, or deliberately influences emotion or other affective phenomena."[11]

Affectiva codes a variety of emotion-related applications, primarily using deep learning techniques. These range from detecting distracted and "risky" drivers on roads to measuring the emotional responses of consumers to advertising. The company has built what they call the world's largest emotion database, made up of over ten million people's expressions from eighty-seven countries.[12] Their monumental collection of videos of people emoting was hand labeled by crowdworkers based primarily in Cairo.[13] Many more companies have now licensed Affectiva's products to develop everything from applications that assess job candidates to analyzing whether students are engaged in class, all by capturing and analyzing their facial expressions and body language.[14]

Beyond the start-up sector, AI giants like Amazon, Microsoft, and IBM have all designed systems for affect and emotion detection. Microsoft offers emotion detection in its Face API, which claims to detect what an individual is feeling across the emotions of "anger, contempt, disgust, fear, happiness, neutral, sadness, and surprise" and asserts that "these emotions are understood to be cross-culturally and universally communicated with particular facial expressions."[15] Amazon's Rekognition tool similarly claims that it can identify "all seven emotions" and "measure how these things change over time, such as constructing a timeline of the emotions of an actor."[16]

But how do these technologies work? Emotion recognition systems grew from the interstices between AI technologies, military priorities, and the behavioral sciences—psychology in particular. They share a similar set of blueprints and founding assumptions: that there is a small number of distinct and universal emotional categories, that we involuntarily reveal these emotions on our faces, and that they can be detected by machines. These articles of faith are so accepted in some fields that it can seem strange even to notice them, let alone question them. They are so ingrained that they have come to constitute "the common view."[17] But if we look at how emotions came to be taxonomized—neatly ordered and labeled—we see that questions are lying in wait at every corner. And a leading figure behind this approach is Paul Ekman.

"The World's Most Famous Face-Reader"

Ekman's research began with a fortunate encounter with Silvan Tomkins, then an established psychologist based at Princeton who had published the first volume of his magnum opus, *Affect Imagery Consciousness*, in 1962.[18] Tomkins's work on affect had a huge influence on Ekman, who devoted much of his career to studying its implications. One aspect in particular played an outsized role: the idea that if affect was an innate set of evolutionary responses, they would be universal and so recognizable across cultures. This desire for universality has an important bearing on why these theories are widely applied in AI emotion recognition systems today: it offered a small set of principles that could be applied everywhere, a simplification of complexity that was easily replicable.

In the introduction to *Affect Imagery Consciousness*, Tomkins framed his theory of biologically based universal affects as one addressing an acute crisis of human sovereignty. He was challenging the development of behaviorism and psychoanalysis, two schools of thought that he believed treated consciousness as a mere by-product of—and in service to—other forces. He noted that human consciousness had "been challenged and reduced again and again, first by Copernicus"—who displaced man from the center of the universe—"then by Darwin"—whose theory of evolution shattered the idea that humans were created in the image of a Christian God—"and most of all by Freud"—who decentered human consciousness and reason as the driving force behind our motivations.[19] Tomkins continued, "The paradox of maximal control over nature and minimal control over human nature is in part a derivative of the neglect of the role of consciousness as a control mechanism."[20] To put it simply, *consciousness tells us little about why we feel and act the way we do.* This is a critical claim for all sorts of later applications of affect theory, which stress the inability of humans to recognize both the feeling and the expression of affects. If we as humans are incapable of truly detecting what we are feeling, then perhaps AI systems can do it for us?

Tomkins's theory of affects was his way to address the problem of human motivation. He argued that motivation was governed by two systems: affects and drives. Tomkins contended that drives tend to be closely associated with immediate biological needs such as hunger and thirst.[21] They are instrumental; the pain of hunger can be remedied with food. But the primary system governing human motivation and behavior is that of affects, involving positive and negative *feelings.* Affects, which play the most important role in human motivation, amplify drive signals, but they are much more complex. For example, it is difficult to know the precise reason or causes that lead a baby to cry, expressing the distress-anguish affect. The baby might be "hungry or cold or wet or in pain or [crying] because of a high temperature."[22] Similarly, there are a number of ways that this affective feeling can be managed: "Crying can be stopped by feeding, cuddling, making the room warmer, making it colder, taking the diaper pin out of his skin and so on."[23]

Tomkins concludes, "The price that is paid for this flexibility is ambiguity and error. The individual may or may not correctly identify the 'cause' of his fear or joy and may or may not learn to reduce his fear or maintain or recapture his joy. In this respect the affect system is not as simple a signal system as the drive system."[24] Affects, unlike drives, are not strictly instrumental; they have a high degree of independence from stimuli and objects, meaning that we often may not know why we feel angry, afraid, or happy.[25]

All of this ambiguity might suggest that the complexities of affects are impossible to untangle. How can we know anything about a system where the connections between cause and effect, stimulus and response, are so tenuous and uncertain? Tomkins proposed an answer: "The primary affects . . . seem to be innately related in a one-to-one fashion with an organ system which is extraordinarily visible." Namely, the face.[26] He found precedents for this emphasis on facial expression in two works published in the nineteenth century: Charles Darwin's *The Expression of the Emotions in Man and Animals* (1872) and an obscure volume by the French neurologist Guillaume-Benjamin-Amand Duchenne de Boulogne, *Mécanisme de la physionomie humaine ou Analyse électro-physiologique de l'expression des passions applicable à la pratique des arts plastiques* (1862).[27]

Tomkins assumed that the facial display of affects was a human universal. "Affects," Tomkins believed, "are sets of muscle, vascular, and glandular responses located in the face and also widely distributed through the body, which generate sensory feedback. . . . These organized sets of responses are triggered at subcortical centers where specific 'programs' for each distinct affect are stored"—a very early use of a computational metaphor for a human system.[28]

But Tomkins acknowledged that the *interpretation* of affective displays depends on individual, social, and cultural factors. He admitted that there were very different "dialects" of facial language in different societies.[29] Even the forefather of affect research raised the possibility that recognizing affect and emotion depends on social and cultural context. The potential conflict between cultural dialects and a biologically based, universal language had enormous implications for the study of facial expression and later forms of emotion recognition. Given that facial expressions are culturally variable, using them to train machine learning systems would inevitably mix together all sorts of different contexts, signals, and expectations.

During the mid-1960s, opportunity knocked at Ekman's door in the form of the Advanced Research Projects Agency (ARPA), a research arm of the Department of Defense. Looking back on this period, he admitted, "It wasn't my idea to do this [affect research]. I was asked—pushed. I didn't even write the research proposal. It was written for me by the man who gave me the money to do it."[30] In 1965, he was researching nonverbal expression in clinical settings and seeking funding to develop a research program at Stanford University. He arranged a meeting in Washington, D.C., with Lee Hough, head of ARPA's behavioral sciences division.[31] Hough was uninterested in how Ekman described his research, but he saw potential in understanding cross-cultural nonverbal communication.[32]

The only problem was that, by Ekman's own admission, he did not know how to do cross-cultural research: "I did not even know what the arguments were, the literature, or the methods."[33] So Ekman understandably decided to drop pursuit of ARPA funding. But Hough insisted, and according to Ekman, he "sat for a day in my office, and wrote the proposal he then funded that allowed me to do the research I am best known for—evidence for the universality of some facial expressions of emotion, and cultural differences in gestures."[34] He got a massive injection of funds from ARPA, roughly one million dollars—the equivalent of more than eight million dollars today.[35]

At the time, Ekman wondered why Hough seemed so eager to fund this research, even over his objections and despite his lack of expertise. It turns out that Hough wanted to distribute his money quickly to avoid suspicion from Senator Frank Church, who had caught Hough using social science research as a cover for acquiring information in Chile that could be used to overthrow its left-wing government under President Salvador Allende.[36] Ekman later concluded that he was just a lucky guy, someone "who could do overseas research that wouldn't get him [Hough] into trouble!"[37] ARPA would be the first in a long line of agencies from defense, intelligence, and law enforcement that would fund both Ekman's career and the field of affect recognition more generally.

With the support of a large grant behind him, Ekman began his first studies to prove universality in facial expression. In general, these studies followed a design that would be copied in early AI labs. He largely duplicated Tomkins's methods, even using Tomkins's photographs to test subjects drawn from Chile, Argentina, Brazil, the United States, and Japan.[38] He relied on asking research participants to simulate the expressions of an emotion, which were then compared with expressions gathered "in the wild," meaning outside of laboratory conditions.[39] Subjects were presented with photographs of posed facial expressions, selected by the designers as exemplifying or expressing a particularly "pure" or intense affect. Subjects were then asked to choose among these affect categories and to label the posed image. The analysis measured the degree to which the labels chosen by subjects correlated with those chosen by the designers.

From the start, the methodology had problems. Ekman's forced choice response format would be later criticized for alerting subjects to the connections that designers had already made between facial expressions and emotions.[40] Further, the fact that these emotions were faked or posed would raise significant concerns about the validity of these results.[41]

Ekman found some cross-cultural agreements using this approach, but his findings were challenged by the anthropologist Ray Birdwhistell, who suggested that this agreement may not reflect innate affect states if they were culturally learned through exposure to such mass media as films, television, or magazines.[42] It was this dispute that compelled Ekman to set out for Papua New Guinea, specifically to study indigenous people in the highlands region. He figured that if people with little contact to Western culture and media could agree with how he had categorized posed affective expressions, then this would provide strong evidence for the universality of his schema.

After Ekman returned from his first attempt to study the Fore people in Papua New Guinea, he devised an alternative approach to prove his theory. He showed his U.S. research subjects a photograph, then asked them to choose one of six affect concepts: happy, fear, disgust-contempt, anger, surprise, and sadness.[43] The results were close enough to subjects from other countries that Ekman believed he could claim that "particular facial behaviors are universally associated with particular emotions."[44]

Affect: From Physiognomy to Photography

The idea that interior states can be reliably inferred from external signs stems in part from the history of physiognomy, which was premised on studying a person's facial features for indications of their character. In the ancient Greek world, Aristotle had believed that "it is possible to judge men's character from their physical appearance . . . for it has been assumed that body and soul are affected together."[45] The Greeks also used physiognomy as an early form of racial classification, applied to "the genus man itself, dividing him into races, in so far as they differ in appearance and in character (for instance Egyptians, Thracians and Scythians)."[46] They presumed a link between body and soul that justified reading a person's interior character based on their exterior appearance.

Physiognomy in Western culture reached a high point during the eighteenth and nineteenth centuries, when it was seen as part of the anatomical sciences. A key figure in this tradition was the Swiss pastor Johann Kaspar Lavater, who wrote *Essays on Physiognomy; For the Promotion of Knowledge and the Love of Mankind*, originally published in German in 1789.[47] Lavater took the approaches of physiognomy and blended them with the latest scientific knowledge. He tried to create a more "objective" comparison of faces by using silhouettes instead of artists' engravings because they were more mechanical and fixed the position of each face into the familiar profile form, allowing for a comparative viewpoint.[48] He believed that bone structure was an underlying connection between physical appearance and character type. If facial expressions were fleeting, skulls offered a more solid material for physiognomic inferences.[49] The measurement of skulls, as we saw in the last chapter, was used to support an emerging nationalism, racism, and xenophobia. This work was infamously elaborated on throughout the nineteenth century by phrenologists like Franz Joseph Gall and Johann Gaspar Spurzheim, as well as in scientific criminology through the work of Cesare Lombroso—all leading into the types of inferential classifications that recur in contemporary AI systems.

But it was the French neurologist Duchenne, described by Ekman as a "marvelously gifted observer," who codified the use of photography and other technical means in the study of human faces.[50] In *Mécanisme de la physionomie humaine*, Duchenne laid important foundations for both Darwin and Ekman, connecting older ideas from physiognomy and phrenology with more modern investigations into physiology and psychology. He replaced vague assertions about character with a more limited investigation into expression and interior mental or emotional states.[51]

Duchenne worked in Paris at the Salpetrière asylum, which housed up to five thousand people with a wide range of diagnoses of mental illness and neurological conditions. Some would become his subjects for distressing experiments, part of the long tradition of medical and technological experimentation on the most vulnerable and those who cannot refuse.[52] Duchenne, who was little known in the scientific community, decided to develop techniques of electrical shocks to stimulate isolated muscle movements in people's faces. His aim was to build a more complete anatomical and physiological understanding of the face. Duchenne used these methods to bridge the new psychological science and the much older study of physiognomic signs, or passions.[53] He relied on the latest photographic techniques, like collodion processing, which allowed for much shorter exposure times, allowing Duchenne to freeze fleeting muscular movements and facial expressions in images.[54]

Even at these very early stages, the faces were never natural or socially occurring human expressions but *simulations* produced by the brute application of electricity to the muscles. Regardless, Duchenne believed that the use of photography and other technical systems would transform the squishy business of representation into something objective and evidentiary, more suitable for scientific study.[55] In his introduction to *The Expression of the Emotions in Man and Animals*, Darwin praised Duchenne's "magnificent photographs" and included reproductions in his own work.[56] Because emotions were temporal, even fleeting occurrences, photography offered the ability to fix, compare, and categorize their visible expres-

Plates from G.-B. Duchenne (de Boulogne),
Mécanisme de la physionomie humaine, ou Analyse
électro-physiologique de l'expression des passions.
Courtesy U.S. National Library of Medicine

sion on the face. Yet Duchenne's images of truth were highly manufactured.

Ekman would follow Duchenne in placing photography at the center of his experimental practice.[57] He believed that slow motion photography was essential to his approach, because many facial expressions operate at the limits of human perception. The aim was to find so-called microexpressions — tiny muscle movements in the face. The duration of microexpressions, in his view, "is so short that they are at the threshold of recognition unless slow motion projection is utilized."[58] In later years Ekman also would insist that anyone could come to learn to recognize microexpressions, with no special training or slow motion capture, in about an hour.[59] But if these expressions are too quick for humans to recognize, how are they to be understood?[60]

One of Ekman's ambitious plans in his early research was to codify a system for detecting and analyzing facial expressions.[61] In 1971, he copublished a description of what he called the Facial Action Scoring Technique (FAST). Relying on posed photographs, the approach used six basic emotional types largely derived from Ekman's intuitions.[62] But FAST soon ran into problems when other scientists were able to produce facial expressions not included in its typology.[63] So Ekman decided to ground his next measurement tool in facial musculature, harkening back to Duchenne's original electroshock studies. Ekman identified roughly forty distinct muscular contractions on the face and called the basic components of each facial expression an Action Unit.[64] After some testing and validation, Ekman and Wallace Friesen published the Facial Action Coding System (FACS) in 1978; the updated editions continue to be widely used.[65] FACS was very labor intensive to use as a measurement tool. Ekman said that it took from seventy-five

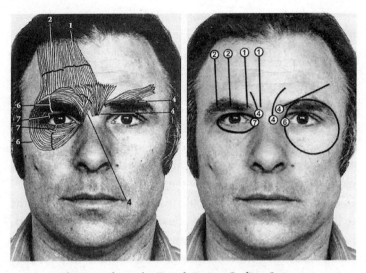

Elements from the Facial Action Coding System.
Source: Paul Ekman and Wallace V. Friesen

to a hundred hours to train users in the FACS methodology and an hour to score a minute of facial footage.[66]

At a conference in the early 1980s, Ekman heard a research presentation that suggested a solution to the intense labor demands of FACS: the use of computers to automate measurement. Although in his memoir Ekman does not mention the researcher who gave the paper, he does state that the system was called Wizard and was developed at Brunel University in London.[67] This is likely Igor Aleksander's early machine learning object-recognition system, WISARD, which had used neural networks at a time when this approach was out of fashion.[68] Some sources report that WISARD was trained on a "database of known football hooligans," anticipating the widespread contemporary use of criminal mug shots to train facial recognition technologies.[69]

Because facial recognition emerged as a foundational application for artificial intelligence in the 1960s, it is not surprising that early researchers working in this field found common cause with Ekman's approach to analyzing faces.[70] Ekman himself claims to have played an active role in driving the automated forms of affect recognition through his old contacts in defense and intelligence agencies from his ARPA funding days. He helped to set up an informal competition between two teams working with FACS data, and this seems to have had lasting impact. Both of those teams have since gone on to feature prominently in the affective computing field. One team was composed of Terry Sejnowski and his student Marian Bartlett, who herself became an important figure in the computer science of emotion recognition and the lead scientist at Emotient, acquired by Apple in 2016.[71] The second team, based in Pittsburgh, was led by the psychologist Jeffrey Cohn of the University of Pittsburgh and the eminent computer vision researcher Takeo Kanade of Carnegie Mellon.[72] These two figures pursued affect recognition over the long term and developed the well-known Cohn-Kanade (CK) emotional expression dataset and its descendants.

Ekman's FACS system provided two things essential for later machine learning applications: a stable, discrete, finite set of labels that humans can use to categorize photographs of faces and a system for producing measurements. It promised to remove the difficult work of representing interior lives away from the purview of artists and novelists and bring it under the umbrella of a rational, knowable, and measurable rubric suitable to laboratories, corporations, and governments.

Capturing Feeling: The Artifice
of Performing Emotions

As work into the use of computers in affect recognition began to take shape, researchers recognized the need for a collection of standardized images to experiment with. A 1992 NSF report coauthored by Ekman recommended that "a readily accessible, multimedia database shared by the diverse facial research community would be an important resource for the resolution and extension of issues concerning facial understanding."[73] Within a year, the Department of Defense would begin funding the FERET program to collect facial photographs, as we saw in chapter 3. By the end of the decade, machine learning researchers had begun to assemble, label, and make public the datasets that drive much of today's machine learning research.

Ekman's FACS guidelines directly shaped the CK dataset.[74] Following Ekman's tradition of posed facial expressions, "subjects were instructed by an experimenter to perform a series of 23 facial displays," which FACS experts then coded, providing labels for the data. The CK dataset allowed laboratories to benchmark their results and compare progress as they built new expression recognition systems.

Other labs and companies worked on parallel projects, creating scores of photo databases. For example, researchers in a lab in Sweden created Karolinska Directed Emotional Faces. This database is composed of images of individuals portraying posed emotional expressions corresponding to Ekman's categories.[75] They make their faces into the shapes that accord with six basic emotional states. When looking at these training sets, it is difficult to not be struck by how extreme they are: *Incredible surprise! Abundant joy! Paralyzing fear!* These subjects are literally making machine-readable emotion.

As the field grew in scale and complexity, so did the types

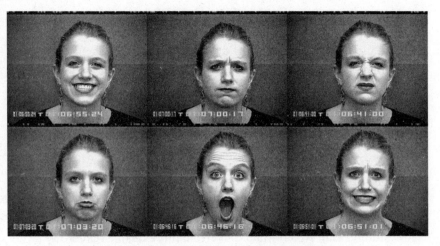

Facial expressions from the Cohn-Kanade dataset: joy, anger, disgust, sadness, surprise, fear. Posed images from T. Kanade et al., *Yearbook of Physical Anthropology* (2000). © Cohn & Kanade

of photographs used in affect recognition. Researchers began using the FACS system to label data generated not from posed expressions but rather from spontaneous facial expressions, sometimes gathered outside of laboratory conditions. For example, a decade after the hugely successful release of the CK dataset, a group of researchers released a second generation, the Extended Cohn-Kanade (CK+) Dataset.[76] CK+ included the usual range of posed expressions but also began to include so-called non-posed or spontaneous expressions taken from videos where subjects made unprompted facial expressions.

By 2009, Affectiva emerged from the MIT Media Lab with the aim of capturing "naturalistic and spontaneous facial expressions" in real-life settings.[77] The company collected data by allowing users to opt into a system that would record their faces using a webcam as they watched a series of commercials. These images would then be hand-labeled using custom software by coders trained in Ekman's FACS.[78] But here we find another problem of circularity. FACS was developed from Ekman's substantial archive of posed photographs.[79] Even when images are gathered in naturalistic settings, they are commonly classified according to a scheme derived from posed images.

Ekman's work became a profound and wide-ranging influence on everything from lie detection software to computer vision. *The New York Times* described Ekman as "the world's most famous face reader," and *Time* named him one of the one hundred most influential people in the world. He would eventually consult with clients as disparate as the Dalai Lama, the FBI, the CIA, the Secret Service, and even the animation studio Pixar, which wanted to create more lifelike renderings of cartoon faces.[80] His ideas became part of popular culture, included in best sellers like Malcolm Gladwell's *Blink* and a television drama, *Lie to Me,* on which Ekman was a consultant for the lead character's role, apparently loosely based on him.[81]

His business also prospered: Ekman sold techniques of deception detection to security agencies such as the Transportation Security Administration, which used them in the development of the Screening of Passengers by Observation Techniques (SPOT) program. SPOT was used to monitor facial expressions of air travelers in the years following the September 11 attacks, attempting to "automatically" detect terrorists. The system uses a set of ninety-four criteria, all of which are allegedly signs of stress, fear, or deception. But looking for these responses meant that some groups are immediately disadvantaged. Anyone who was stressed, was uncomfortable under questioning, or had had negative experiences with police and border guards could score higher. This produced its own forms of racial profiling. The SPOT program has been criticized by the Government Accountability Office and civil liberties groups for its lack of scientific methodology and, despite its nine-hundred-million-dollar price tag, producing no clear successes.[82]

The Many Critiques of Ekman's Theories

As Ekman's fame grew, so did the skepticism of his work, with critiques emerging from a number of fields. An early critic was the cultural anthropologist Margaret Mead, who debated Ekman on the question of the universality of emotions in the late 1960s, resulting in fierce exchanges not only between Mead and Ekman but also among other anthropologists critical of Ekman's idea of absolute universality.[83] Mead was unconvinced by Ekman's belief in universal, biological determinants of behavior rather than considering cultural factors.[84] In particular, Ekman tended to collapse emotions into an oversimplified, mutually exclusive binary: either emotions were universal or they were not. Critics like Mead pointed out that more nuanced positions were possible.[85] Mead took a middle ground, emphasizing that there was no inherent contradiction between "the possibility that human beings may share a core of innate behaviors . . . and the idea that emotional expressions could, *at the same time,* be highly-conditioned by cultural factors."[86]

More scientists from different fields joined the chorus over the decades. In more recent years, the psychologists James Russell and José-Miguel Fernández-Dols have shown that the most basic aspects of the science remain unsolved: "The most fundamental questions, such as whether 'facial expressions of emotion' in fact express emotions, remain subjects of great controversy."[87] Social scientists Maria Gendron and Lisa Feldman Barrett have pointed to the specific dangers of Ekman's theories being used by the AI industry because the automated detection of facial expressions does not reliably indicate an internal mental state.[88] As Barrett observes, "Companies can say whatever they want, but the data are clear. They can detect a scowl, but that's not the same thing as detecting anger."[89]

More troubling still is that in the field of the study of emotions, there is no consensus among researchers about what an emotion actually is. What emotions are, how they are formulated within us and expressed, what their physiological or neurobiological functions could be, their relation to stimuli, even how to define them—all of this in its entirety remains stubbornly unsettled.[90]

Perhaps the foremost critic of Ekman's theory of emotions is the historian of science Ruth Leys. In *The Ascent of Affect* she thoroughly pulls apart "the implications of the fundamental physiognomic assumption underlying Ekman's work . . . namely, the idea that a distinction can be strictly maintained between authentic and artificial expressions of emotion based on differences between the faces we make when we are alone and those we make when we are with others."[91] Leys sees a fundamental circularity in Ekman's method. First, the posed or simulated photographs he used were assumed to express a set of basic affective states, "already free of cultural influence."[92] Then, these photographs were used to elicit labels from different populations to demonstrate the universality of facial expressions. Leys points out the serious problem: Ekman assumed that "the facial expressions in the photographs he employed in his experiments must have been free of cultural taint because they were universally recognized. At the same time, he suggested that those facial expressions were universally recognized because they were free of cultural taint."[93] The approach is fundamentally recursive.[94]

Other problems became clear as Ekman's ideas were implemented in technical systems. As we've seen, many datasets underlying the field are based on actors simulating emotional states, performing for the camera. That means that AI systems are trained to recognize faked expressions of feeling. Although AI systems claim to have access to ground truth about natural interior states, they are trained on material that is inescapably constructed. Even for images that are captured of people responding to commercials or films, those people are aware they are being watched, which can change their responses.

The difficulty in automating the connection between facial movements and basic emotional categories leads to the larger question of whether emotions can be adequately grouped into a small number of discrete categories at all.[95] This view can be traced back to Tomkins, who argued that "each kind of emotion can be identified by a more or less unique signature response within the body."[96] But there is very little consistent evidence of this. Psychologists have conducted multiple reviews of the published evidence, which has failed to find associations among measurable responses to the emotional states that they assume to exist.[97] Finally, there is the stubborn issue that facial expressions may indicate little about our honest interior states, as anyone who has smiled without feeling truly happy can confirm.[98]

None of these serious questions about the basis for Ekman's claims have stopped his work from attaining a privileged role in current AI applications. Hundreds of papers cite Ekman's view of interpretable facial expressions as though it were unproblematic fact, despite decades of scientific controversy. Few computer scientists have even acknowledged this literature of uncertainty. The affective computing researcher Arvid Kappas, for example, directly names the lack of basic scientific consensus: "We know too little regarding the complex social modulators of facial and possibly other expressive activity in such situations to be able to measure emotional state reliably from expressive behavior. *This is not an engineering problem that could be solved with a better algorithm.*[99] Unlike many in the field who confidently support affect recognition, Kappas questions the belief that it's a good idea for computers to be trying to sense emotions at all.[100]

The more time researchers from other backgrounds spend examining Ekman's work, the stronger the evidence against it grows. In 2019, Lisa Feldman Barrett led a research team that conducted a wide-ranging review of the literature on inferring emotions from facial expressions. They concluded firmly that facial expressions are far from indisputable and are "not 'fingerprints' or diagnostic displays" that reliably signal emotional states, let alone across cultures and contexts. Based on all the current evidence, the team observed, "It is not possible to confidently infer happiness from a smile, anger from a scowl, or sadness from a frown, as much of current technology tries to do when applying what are mistakenly believed to be the scientific facts."[101]

Barrett's team was critical of AI companies claiming to be able to automate the inference of emotion: "Technology companies, for example, are spending millions of research dollars to build devices to read emotions from faces, erroneously taking the common view as a fact that has strong scientific support. . . . In fact, our review of the scientific evidence indicates that very little is known about how and why certain facial movements express instances of emotion, particularly at a level of detail sufficient for such conclusions to be used in important, real-world applications."[102]

Why, with so many critiques, has the approach of "reading emotions" from the face endured? By analyzing the history of these ideas, we can begin to see how military research funding, policing priorities, and profit motives have shaped the field. Since the 1960s, driven by significant Department of

Defense funding, multiple systems have been developed that are increasingly accurate at measuring movements on faces. Once the theory emerged that it is possible to assess internal states by measuring facial movements and the technology was developed to measure them, people willingly adopted the underlying premise. The theory fit what the tools could do. Ekman's theories seemed ideal for the emerging field of computer vision because they could be automated at scale.

There are powerful institutional and corporate investments in the validity of Ekman's theories and methodologies. Recognizing that emotions are not easily classified, or that they're not reliably detectable from facial expressions, could undermine an expanding industry. In the AI field, Ekman is commonly cited as though the issue was settled, before directly proceeding into engineering challenges. The more complex issues of context, conditioning, relationality, and cultural factors are hard to reconcile with the current disciplinary approaches of computer science or the ambitions of the commercial tech sector. So Ekman's basic emotional categories became standard. More subtle approaches, like Mead's middle ground, were largely overlooked. The focus has been on increasing the accuracy rates of AI systems rather than on addressing the bigger questions about the many ways we experience, show, and hide emotion and how we interpret the facial expressions of others.

As Barrett writes, "Many of the most influential models in our science assume that emotions are biological categories imposed by nature, so that emotion categories are *recognized*, rather than constructed, by the human mind."[103] AI systems for emotion detection are premised on this idea. Recognition might be the wrong framework entirely when thinking about emotions because recognition assumes that emotional categories are givens, rather than emergent and relational.

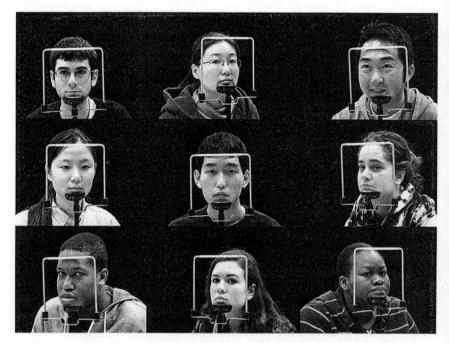

Columbia Gaze Dataset. From Brian A. Smith et al., "Gaze Locking:
Passive Eye Contact Detection for Human-Object Interaction,"
ACM Symposium on User Interface Software and Technology (UIST),
October 2013, 271–80. Courtesy of Brian A. Smith

The Politics of Faces

Instead of trying to build more systems that can group expressions into machine-readable categories, we should question the
origins of those categories themselves, as well as their social
and political consequences. Already, affect recognition tools
are being deployed in political attacks. For example, a conservative blog claimed to create a "virtual polygraph system" to
assess videos of Congresswoman Ilhan Abdullahi Omar.[104] By
using face and speech analytics from Amazon's Rekognition,

XRVision Sentinel AI, and IBM Watson, the blogger claimed that Omar's analytic lie score exceeded her "truth baseline" and that she was registering high on stress, contempt, and nervousness. Several conservative media outlets ran with the story, claiming that Omar is a "pathological liar" and a security threat to the nation.[105]

It's known that these systems flag the speech affects of women differently from men, particularly Black women. As we saw in chapter 3, the construction of the "average" from unrepresentative training data is epistemologically suspect from the outset, with clear racial biases. A study conducted at the University of Maryland has shown that some facial recognition software interprets Black faces as having more negative emotions than white faces, particularly registering them as angrier and more contemptuous, even controlling for their degree of smiling.[106]

This is the danger of affect recognition tools. As we've seen, they take us back to the phrenological past, where spurious claims were made, allowed to stand, and deployed to support existing systems of power. The decades of scientific controversies around the idea of inferring distinct emotions from human faces underscores a central point: the one-size-fits-all recognition model is not the right metaphor for identifying emotional states. Emotions are complex, and they develop and change in relation to our families, friends, cultures, and histories, all the manifold contexts that live outside of the AI frame. In many cases, emotion detection systems do not do what they claim. Rather than directly measuring people's interior mental states, they merely statistically optimize correlations of certain physical characteristics among facial images. The scientific foundations of automated emotion detection are in question, yet a new generation of affect tools is already making inferences across a growing range of high-stakes contexts from policing to hiring.

Even though evidence now points to the unreliability of affect detection, companies continue to seek out new sources to mine for facial imagery, vying for the leading market share of a sector that promises billions in profits. Barrett's systemic review of the research behind inferring emotion from people's faces concludes on a damning note: "More generally, tech companies may well be asking a question that is fundamentally wrong. Efforts to simply 'read out' people's internal states from an analysis of their facial movements alone, without considering various aspects of context, are at best incomplete and at worst entirely lack validity, no matter how sophisticated the computational algorithms. . . . It is premature to use this technology to reach conclusions about what people feel on the basis of their facial movements."[107]

Until we resist the desire to automate affect recognition, we run the risk of job applicants being judged unfairly because their microexpressions do not match other employees, students receiving poorer grades than their peers because their faces indicate a lack of enthusiasm, and customers being detained because an AI system flagged them as likely shoplifters based on their facial cues.[108] These are the people who will bear the costs of systems that are not just technically imperfect but based on questionable methodologies.

The areas of life in which these systems are operating are expanding as rapidly as labs and corporations can create new markets for them. Yet they all rely on a narrow understanding of emotions—grown from Ekman's initial set of anger, happiness, surprise, disgust, sadness, and fear—to stand in for the infinite universe of human feeling and expression across space and time. This takes us back to the profound limitations of capturing the complexities of the world in a single classificatory schema. It returns us to the same problem we have seen repeated: the desire to oversimplify what is stubbornly complex so that it can be easily computed, and packaged for the market. AI systems are seeking to extract the mutable, private, divergent experiences of our corporeal selves, but the result is a cartoon sketch that cannot capture the nuances of emotional experience in the world.

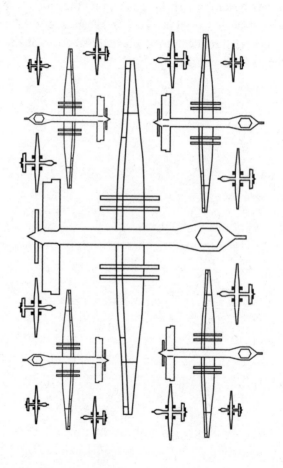

6

State

I'm sitting in front of an air-gapped laptop on the tenth floor of a warehouse building in New York. On the screen is a software program normally used for digital forensics, a tool for investigating evidence and validating information held on hard drives. I'm here to research an archive that contains some of the most specific details about how machine learning began to be used in the intelligence sector, as led by some of the wealthiest governments in the world. This is the Snowden archive: all the documents, PowerPoint presentations, internal memos, newsletters, and technical manuals that former NSA contractor and whistleblower Edward Snowden leaked in 2013. Each page is marked with a header noting different forms of classification. TOP SECRET // SI // ORCON // NOFORN.[1] Each is a warning and a designation.

The filmmaker Laura Poitras first gave me access to this archive in 2014. It was overwhelming to read: the archive held well over a decade of intelligence thinking and communication, including internal documents of the National Security Agency in the United States and the Government Communication Headquarters in the United Kingdom, and the international network of the Five Eyes.[2] This knowledge was strictly off-limits to those without high-level clearance. It was part of the "classified empire" of information, once estimated to be growing five times faster than publicly accessible knowledge but now is anyone's guess.[3] The Snowden archive captures the years when the collection of data metastasized: when phones, browsers, social media platforms, and email all became data sources for the state. The documents reveal how the intelligence community contributed to the development of many of the techniques we now refer to as artificial intelligence.

The Snowden archive reveals a parallel AI sector, one developed in secrecy. The methods share many similarities, but there are striking differences in terms of the reach, the objectives, and the result. Gone are any rhetorical constructs justifying extraction and capture: every software system is simply described as something to be owned, to be defeated; all data platforms are fair game, and very little is designated as protected. One NSA PowerPoint deck outlines TREASUREMAP, a program designed to build a near real-time, interactive map of the internet.[4] It claims to track the location and owner of any connected computer, mobile device, or router: "Map the entire internet—any device, anywhere, all the time," the slide boasts. A few slides on "TREASUREMAP as an Enabler" offers up a layer-cake image of signals analysis. Above the geographical layer and the network layer is the "cyber persona layer"— quaintly represented on the slide by jellybean-era iMacs and Nokia feature phones—and then the "persona layer" of personal connections. This is meant to depict all people who use connected devices around the world, in a "300,000-foot view of the internet." It also looks remarkably like the work of social network mapping and manipulation companies like Cambridge Analytica.

The Snowden documents were released in 2013, but they still read like the AI marketing brochures of today. If

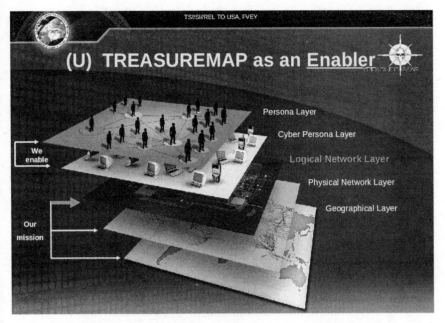

TREASUREMAP as an Enabler. Snowden archive

TREASUREMAP was a precursor to Facebook's God's-eye network view, then the program called FOXACID is reminiscent of Amazon Ring for a home computer: recording everyday activity.[5] "If we can get the target to visit us in some sort of browser, we can probably own them," the slide explains.[6] Once individuals have been tempted to click on a spam email or visit a captured website, the NSA drops files through a browser that will permanently live on their device, quietly reporting everything they do back to base. One slide describes how analysts "deploy very targeted emails" that require "a level of guilty knowledge" about the target.[7] The restrictions on the NSA gathering that guilty knowledge (when it comes to data from American citizens, at least) are rarely discussed. One document notes that the agency was working on multiple fronts to "aggressively pursue legal authorities and a policy framework mapped more fully to the information age."[8] In other words, change the laws to fit the tools, not the other way around.

The U.S. intelligence agencies are the old guards of big data. Along with the Defense Advanced Research Projects Agency, they have been major drivers of AI research since the 1950s. As the historian of science Paul Edwards describes in *The Closed World*, military research agencies actively shaped the emerging field that would come to be known as AI from its earliest days.[9] The Office of Naval Research, for example, partly funded the first Summer Research Project on Artificial Intelligence at Dartmouth College in 1956.[10] The field of AI has always been strongly guided by military support and often military priorities, long before it was clear that AI could be practical at scale. As Edwards notes:

> As the project with the least immediate utility and the farthest-reaching ambitions, AI came to rely unusually heavily on ARPA funding. As a result, ARPA became the primary patron for the first twenty years of AI research. Former director Robert Sproull proudly concluded that "a whole generation of computer experts got their start from DARPA funding" and that "all the ideas that are going into the fifth-generation [advanced computing] project [of the mid-1980s] — artificial intelligence, parallel computing, speech understanding, natural-languages programming — ultimately started from DARPA-funded research."[11]

The military priorities of command and control, automation, and surveillance profoundly shaped what AI was to become. The tools and approaches that came out of DARPA funding have marked the field, including computer vision, automatic translation, and autonomous vehicles. But these technical methods have deeper implications. Infused into the overall logics of AI are certain kinds of classificatory thinking—from explicitly battlefield-oriented notions such as target, asset, and anomaly detection to subtler categories of high, medium, and low risk. Concepts of constant situational awareness and targeting would drive AI research for decades, creating epistemological frameworks that would inform both industry and academia.

From the point of view of the state, the turn to big data and machine learning expanded the modes of information extraction and informed a social theory of how people can be tracked and understood: *you shall know them by their metadata.* Who is texted, which locations are visited, what is read, when devices spring into action and for what reason—these molecular actions became a vision of threat identification and assessment, guilt or innocence. Harvesting and measuring large aggregates of data at a distance became the preferred way to develop alleged insights into groups and communities as well as assessments of potential targets for killing. The NSA and GCHQ are not unique—China, Russia, Israel, Syria, and many other countries have similar agencies. There are many systems of sovereign surveillance and control, a multitude of war machines that never wind down. The Snowden archive underscores how state and corporate actors collaborate in order to produce what Achille Mbembe calls "infrastructural warfare."[12]

But the relationship between national militaries and the AI industry has expanded beyond security contexts. Technologies once only available to intelligence agencies—that were *extralegal* by design—have filtered down to the state's municipal arms: government and law enforcement agencies. While the NSA has been a focus for privacy concerns, less attention is given to the growing commercial surveillance sector, which aggressively markets its tools and platforms to police departments and public agencies. The AI industry is simultaneously challenging and reshaping the traditional role of states while also being used to shore up and expand older forms of geopolitical power. Algorithmic governance is both part of and exceeds traditional state governance. To paraphrase the theorist Benjamin Bratton, the state is taking on the armature of a machine because the machines have already taken on the roles and register of the state.[13]

Making the Third Offset

The story of the internet's creation has been centered around U.S. military and academic innovation and dominance.[14] But in the space of AI, we see that there is no pure national system. Instead, AI systems operate within a complex interwoven network of multinational and multilateral tools, infrastructures, and labor. Take, for example, a facial recognition system that was rolled out in the streets of Belgrade.[15] The director of police ordered the installation of two thousand cameras in eight hundred locations around the city to capture faces and license plates. The Serbian government signed an agreement with Chinese telecommunications giant Huawei to provide the video surveillance, 4G network support, and unified data and command centers. Such deals are common. Local systems are often hybrids, with infrastructure from China, India, the United States, and elsewhere, with porous boundaries, different security protocols, and potential data backdoors.

But the rhetoric around artificial intelligence is much starker: we are repeatedly told that we are in an AI war. The dominant objects of concern are the supernational efforts of the United States and China, with regular reminders that China has stated its commitment to be the global leader in AI.[16] The data practices of China's leading tech companies, including Alibaba, Huawei, Tencent, and ByteDance, are often framed as direct Chinese state policy and thus seen as inherently more threatening than U.S. private actors such as Amazon and Facebook, even though the lines between state and corporate imperatives and incentives are complexly intertwined. Yet the language of war is more than just the usual articulation of xenophobia, mutual suspicion, international espionage, and network hacking. As media scholars such as Wendy Chun and Tung-Hui Hu have noted, the liberal vision of global digital citizens engaging as equals in the abstract space of networks has shifted toward a paranoid vision of defending a national cloud against the racialized enemy.[17] The specter of the foreign threat works to assert a kind of sovereign power over AI and to redraw the locus of power of tech companies (which are transnational in infrastructure and influence) back within the bounds of the nation-state.

Yet the nationalized race for technological superiority is both rhetorical and real at the same time, creating the dynamics for geopolitical competition across and within commercial and military sectors, increasingly blurring the lines between the two. The dual use of AI applications in both civilian and military domains has also produced strong incentives for close collaboration and funding.[18] In the United States, we can see how this became an explicit strategy: to seek national control and international dominance of AI in order to secure military and corporate advantage.

The latest iteration of this strategy emerged under Ash Carter, who served as U.S. secretary of defense from 2015 to 2017. Carter played a significant role in bringing Silicon Valley into closer relationship to the military, convincing tech companies that national security and foreign policy depended on American dominance of AI.[19] He called this the Third Offset strategy. An offset is generally understood as a way of compensating for an underlying military disadvantage by changing the conditions, or as former secretary of defense Harold Brown stated in 1981, "Technology can be a force multiplier, a resource that can be used to help offset numerical advantages of an adversary. Superior technology is one very effective way to balance military capabilities other than by matching an adversary tank-for-tank or soldier-for-soldier."[20]

The First Offset is commonly understood as the use of nuclear weapons in the 1950s.[21] The Second was the expansion of covert, logistical, and conventional weapons in the 1970s and 1980s. The Third, according to Carter, should be a combination of AI, computational warfare, and robots.[22] But unlike the NSA, which already had robust surveillance capabilities, the U.S. military lacked the AI resources, expertise, and infrastructure of America's leading technology companies.[23] In 2014, Deputy Defense Secretary Robert Work outlined the Third Offset as an attempt to "exploit all the advances in artificial intelligence and autonomy."[24]

To build AI war machines, the Department of Defense would need gigantic extractive infrastructures. Yet in order to gain access to highly paid engineering labor and sophisticated development platforms, partnering with industry was necessary. The NSA had paved the way with systems like PRISM, both working with and secretly infiltrating telecommunications and technology companies.[25] But these more covert approaches faced renewed political pushback after the Snowden disclosures. Congress passed the USA Freedom Act in 2015, which introduced some limitations on the NSA's access to real-time data from Silicon Valley. Yet the possibility for a larger military-industrial complex around data and AI remained tantalizingly close. Silicon Valley had already built and monetized the logics and infrastructures of AI required to drive a new offset. But first the tech sector had to be convinced that partnering on creating the infrastructure of warfare would be worth it without alienating their employees and deepening public mistrust.

Enter Project Maven

In April 2017, the Department of Defense published a memo announcing the Algorithmic Warfare Cross-Functional Team, code-named Project Maven.[26] "The Department of Defense must integrate artificial intelligence and machine learning more effectively across operations to maintain advantages over increasingly capable adversaries and competitors," wrote the deputy defense secretary.[27] The goal of the program was to get the best possible algorithmic systems into the battlefield quickly, even when they were just 80 percent complete.[28] It was part of a much bigger plan, the Joint Enterprise Defense Infrastructure cloud project—or JEDI—an enormous redesign of the entire IT infrastructure of the Defense Department, from the Pentagon to field-level support. Project Maven was a small piece of this larger picture, and the aim was to create an AI system that would allow analysts to select a target and then see every existing clip of drone footage that featured the same person or vehicle.[29] Ultimately, the Defense Department wanted an automated search engine of drone videos to detect and track enemy combatants.

The technical platforms and machine learning skills needed for Project Maven were centered in the commercial tech sector. The Defense Department decided to pay tech companies to analyze military data collected from satellites and battlefield drones in places where U.S. domestic privacy laws did not apply. This would align military and U.S. tech sector financial interests around AI without directly triggering constitutional privacy tripwires, as the National Security Agency

The official seal of the Algorithmic Warfare Cross-Functional
Team, code-named Project Maven. The Latin motto translates as
"Our job is to help." Produced by U.S. Department of Defense

had done. A bidding war began among the technology companies that wanted the Maven contract, including Amazon,
Microsoft, and Google.

The first Project Maven contract went to Google. Under
the agreement, the Pentagon would use Google's TensorFlow
AI infrastructure to comb through drone footage and detect
objects and individuals as they moved between locations.[30]
Fei-Fei Li, then chief scientist of AI/ML at Google, was already
an expert in building object recognition datasets, given her experience creating ImageNet and using satellite data to detect
and analyze cars.[31] But she was adamant that the project should
be kept secret. "Avoid at ALL COSTS any mention or implication of AI," Li wrote in an email to Google colleagues that was
later leaked. "Weaponized AI is probably one of the most sen-

sitized topics of AI—if not THE most. This is red meat to the media to find all ways to damage Google."[32]

But in 2018, Google employees discovered the extent of the company's role in the project. They were furious that their work was being used for warfare purposes, especially after it became known that Project Maven's image identification goals included objects such as vehicles, buildings, and humans.[33] More than 3,100 employees signed a letter of protest stating that Google should not be in the business of war and demanded that the contract be canceled.[34] Under increasing pressure, Google officially ended its work on Project Maven and withdrew from the competition for the Pentagon's ten-billion-dollar JEDI contract. In October that year, Microsoft's president, Brad Smith, announced in a blog post that "we believe in the strong defense of the United States and we want the people who defend it to have access to the nation's best technology, including from Microsoft."[35] The contract ultimately went to Microsoft, which outbid Amazon.[36]

Shortly after the internal uprising, Google released its Artificial Intelligence Principles, which included a section on "AI applications we will not pursue."[37] These included making "weapons or other technologies whose principal purpose or implementation is to cause or directly facilitate injury to people," as well as "technologies that gather or use information for surveillance violating internationally accepted norms."[38] While the turn to AI ethics quelled some internal and external concerns, the enforceability and parameters of ethical restraint were left unclear.[39]

In response, former Google CEO Eric Schmidt characterized the pushback over Project Maven as "a general concern in the tech community of somehow the military-industrial complex using our stuff to kill people incorrectly, if you will."[40] This shift, from the debate over whether to use AI in warfare at all to a debate over whether AI could help to "kill people correctly," was quite strategic.[41] It moved the focus away from the foundational ethics of AI as a military technology toward questions of precision and technical accuracy. But Lucy Suchman argues that the problems with automated warfare go far beyond whether the killing was accurate or "correct."[42] Particularly in the case of object detection, Suchman asks, who is building the training sets and using what data, and how are things labeled as an imminent threat? What kinds of classificatory taxonomies are used to decide what constitutes sufficiently abnormal activity to trigger a legal drone attack? And why should we condone attaching life or death consequences to these unstable and inherently political classifications?[43]

The Maven episode, as well as the AI principles that emerge, points to the deep schisms in the AI industry about the relationship between the military and civilian spheres. The AI war, both real and imagined, instills a politics of fear and insecurity that creates a climate that is used to stifle internal dissent and promote unquestioning support for a nationalist agenda.[44] After the fallout from Maven faded, Google's chief legal officer, Kent Walker, said that the company was pursuing higher security certifications in order to work more closely with the Defense Department. "I want to be clear," he said. "We are a proud American company."[45] Articulating patriotism as policy, tech companies are increasingly expressing strong alignment with the interests of the nation-state, even as their platforms and capacities exceed traditional state governance.

The Outsourced State

The relationship between the state and the AI industry goes well beyond national militaries. The technologies once reserved for war zones and espionage are now used at the local level of government, from welfare agencies to law enforcement. This shift has been propelled by outsourcing key functions of the state to technology contractors. On the surface, this does not seem very different than the usual outsourcing of government functions to the private sector through companies such as Lockheed Martin or Halliburton. But now militarized forms of pattern detection and threat assessment are moving at scale into municipal-level services and institutions.[46] A significant example of this phenomenon is the company named after the magical seeing stones in *Lord of the Rings:* Palantir.

Palantir was established in 2004, cofounded by PayPal billionaire Peter Thiel, who was also an adviser and financial supporter of President Trump. Thiel would later argue in an opinion piece that AI is first and foremost a military technology: "Forget the sci-fi fantasy; what is powerful about actually existing AI is its application to relatively mundane tasks like computer vision and data analysis. Though less uncanny than Frankenstein's monster, these tools are nevertheless valuable to any army—to gain an intelligence advantage, for example. . . . No doubt machine learning tools have civilian uses, too."[47]

While Thiel recognizes the nonmilitary uses of machine learning, he particularly believes in the *in-between space:* where commercial companies produce military-styled tools to be provided to anyone who would like to gain an intelligence advantage and is willing to pay for it. Both he and Palantir's CEO, Alex Karp, describe Palantir as "patriotic," with Karp accusing other technology companies that refuse to work with the military agencies as "borderline craven."[48] In an insightful essay, the writer Moira Weigel studied Karp's university dissertation, which reveals his early intellectual interest in aggression and a belief that "the desire to commit violence is a constant founding fact of human life."[49] Karp's thesis was titled "Aggression in the Life World."

Palantir's original clients were federal military and intelligence agencies, including the Defense Department, National Security Agency, FBI, and CIA.[50] As revealed in an investigation by Mijente, after Trump took the presidency, Palantir's contracts with U.S. agencies totaled more than a billion dollars.[51] But Palantir did not style itself as a typical defense contractor in the mold of Lockheed Martin. It adopted the character of the Silicon Valley start-up, based in Palo Alto and predominantly staffed by young engineers, and it was backed by In-Q-Tel, the venture capital firm funded by the CIA. Beyond its initial intelligence agency clients, Palantir began to work with hedge funds, banks, and corporations like Walmart.[52] But its DNA was shaped working for, and within, the defense community. It deployed the same approaches seen in the Snowden documents, including extracting data across devices and infiltrating networks in order to track and evaluate people and assets. Palantir quickly became a preferred outsourced surveillance provider, including designing the databases and management software to drive the mechanics of deportation for Immigration and Customs Enforcement (ICE).[53]

Palantir's business model is based on a mix of data analysis and pattern detection using machine learning, combined with more generic consulting. Palantir sends engineers into a company, who extract a wide variety of data—emails, call logs, social media, when employees enter and leave buildings, when they book plane tickets, everything the company is prepared to share—then look for patterns and give advice on what to do next. One common approach is to search for current or potential so-called bad actors, disgruntled employees who may leak information or defraud the company. The underlying worldview built into Palantir's tools is reminiscent of the NSA: collect everything, then look for anomalies in the data. However, while the NSA's tools are built to surveil and target enemies of the state, in either conventional or covert warfare, Palantir's approach has been directed against civilians. As described in a major investigation by Bloomberg in 2018, Palantir is "an intelligence platform designed for the global War on Terror" that is now "weaponized against ordinary Americans at home": "Palantir cut its teeth working for the Pentagon and the CIA in Afghanistan and Iraq. . . . The U.S. Department of Health and Human Services uses Palantir to detect Medicare fraud. The FBI uses it in criminal probes. The Department of Homeland Security deploys it to screen air travelers and keep tabs on immigrants."[54]

Soon, keeping tabs on undocumented workers evolved into capturing and deporting people at schools and places of work. In furtherance of this objective, Palantir produced a phone app called FALCON, which functions as a vast dragnet, gathering data from multiple law enforcement and public databases that list people's immigration histories, family relationships, employment information, and school details. In 2018, ICE agents used FALCON to guide their raid of almost a hundred 7-Elevens across the United States in what was called "the largest operation against a single employer in the Trump era."[55]

Despite Palantir's efforts to maintain secrecy about what it builds or how its systems work, its patent applications give us some insight into the company's approach to AI for deportation. In an application innocuously entitled *Database systems and user interfaces for dynamic and interactive mobile image analysis and identification,* Palantir brags about the app's ability to photograph people in short-time-frame encounters and, regardless of whether they are under suspicion or not, to run their image against all available databases. In essence, the system uses facial recognition and back-end processing to create a framework on which to base any arrest or deportation.

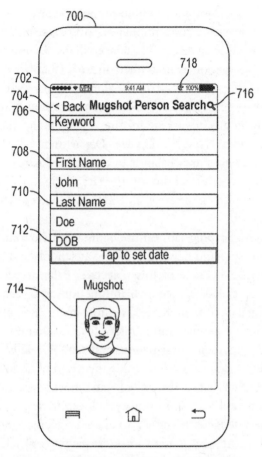

An image from Palantir's patent US10339416B2.
Courtesy U.S. Patent and Trademark Office

While Palantir's systems have structural similarities to those at the NSA, they have devolved to a local community level, to be sold to supermarket chains and local law enforcement alike. This represents a shift away from traditional policing toward the goals more associated with military intelligence infrastructures. As law professor Andrew Ferguson explains,

"We are moving to a state where prosecutors and police are going to say 'the algorithm told me to do it, so I did, I had no idea what I was doing.' And this will be happening at a widespread level with very little oversight."[56]

The sociologist Sarah Brayne was one of the first scholars to observe directly how Palantir's data platforms are used in situ, specifically by the Los Angeles Police Department. After more than two years of riding along with police on patrols, watching them at their desks, and conducting multiple interviews, Brayne concluded that in some domains these tools merely amplify prior police practices but that in other ways they are transforming the process of surveillance entirely. In short, police are turning into intelligence agents:

> The shift from traditional to big data surveillance is associated with a migration of law enforcement operations toward intelligence activities. The basic distinction between law enforcement and intelligence is as follows: law enforcement typically becomes involved once a criminal incident has occurred. Legally, the police cannot undertake a search and gather personal information until there is probable cause. Intelligence, by contrast, is fundamentally predictive. Intelligence activities involve gathering data; identifying suspicious patterns, locations, activity, and individuals; and preemptively intervening based on the intelligence acquired.[57]

Although everyone is subject to these types of surveillance, some people are more likely to be subjected to it than others: immigrants, the undocumented, the poor, and communities of color. As Brayne observed in her study, the use of Palantir's software reproduces inequality, making those in predominantly poor, Black, and Latinx neighborhoods subject to even greater surveillance. Palantir's point system lends an aura of objectivity: it's "just math," in the words of one police officer. But it creates a reinforcing loop of logic.[58] Brayne writes:

> Despite the stated intent of the point system to avoid legally contestable bias in police practices, it hides both intentional and unintentional bias in policing and creates a self-perpetuating cycle: if individuals have a high point value, they are under heightened surveillance and therefore have a greater likelihood of being stopped, further increasing their point value. Such practices hinder the ability of individuals already in the criminal justice system from being further drawn into the surveillance net, while obscuring the role of enforcement in shaping risk scores.[59]

The machine learning approaches of Palantir and its ilk can lead to a feedback loop, where those included in a criminal justice database are more likely to be surveilled and thus more likely to have more information about them included, which justifies further police scrutiny.[60] Inequity is not only deepened but tech-washed, justified by the systems that appear immune to error yet are, in fact, intensifying the problems of overpolicing and racially biased surveillance.[61] The intelligence models that began in national government agencies have now become part of the policing of local neighborhoods. The NSA-ification of police departments exacerbates historical inequality and radically transforms and expands the practices of police work.

Despite the massive expansion of government contracts for AI systems, little attention has been given to the question of whether private vendors of these technologies should be legally accountable for the harms produced when governments use their systems. Given how often governments are turning to contractors to provide the algorithmic architectures for state decision-making, be it policing or welfare systems, there is a case that technology contractors like Palantir should be liable for discrimination and other violations. Currently most states attempt to disclaim any responsibility for problems created by the AI systems they procure, with the argument that "we cannot be responsible for something we don't understand." This means that commercial algorithmic systems are contributing to the process of government decision making without meaningful mechanisms of accountability. With the legal scholar Jason Schultz, I've argued that developers of AI systems that directly influence government decisions should be found to be state actors for purposes of constitutional liability in certain contexts.[62] That is, they could be found legally liable for harms in the same way that states can be. Until then, vendors and contractors have little incentive to ensure that their systems aren't reinforcing historical harms or creating entirely new ones.[63]

Another example of this phenomenon is Vigilant Solutions, established in 2005. The company works on the basis of a single premise: take surveillance tools that might require judicial oversight if operated by governments and turn them into a thriving private enterprise outside constitutional privacy limits. Vigilant began its venture in multiple cities across the United States by installing automatic license-plate recognition (ALPR) cameras, placing them everywhere from cars to light poles, parking lots to apartment buildings. This array of networked cameras photographs every passing car, storing license plate images in a massive perpetual database. Vigilant then sells access to that database to the police, private investigators, banks, insurance companies, and others who want access to it. If police officers want to track a car across the entire state and mark every place it has been, Vigilant can show them. Likewise, if a bank wanted to repossess a car, Vigilant could reveal where it was, for a price.

California-based Vigilant markets itself as "one of those trusted crime fighting tools to help law enforcement develop leads and solve crimes faster," and it has partnered with a range of governments in Texas, California, and Georgia to provide their police with a suite of ALPR systems to use on patrol, along with access to Vigilant's database.[64] In return, the local governments provide Vigilant with records of outstanding arrest warrants and overdue court fees. Any license plates flagged to match those associated with outstanding fines in the database are fed into police officers' mobile systems, altering them to pull these drivers over. Drivers are then given two options: pay the outstanding fine on the spot or be arrested. On top of taking a 25 percent surcharge, Vigilant keeps records of every license plate reading, extracting that data to add to its massive databases.

Vigilant signed a significant contract with ICE that gave the agency access to five billion records of license plates gathered by private businesses, as well as 1.5 billion data points contributed by eighty local law enforcement agencies across the United States—including information on where people live and work. That data can stem from informal arrangements between local police and ICE and may already violate state data-sharing laws. ICE's own privacy policy limits data collection near "sensitive locations" like schools, churches, and protests. But in this case, ICE doesn't collect the data or maintain the database—the agency simply buys access to Vigilant's systems, which has far fewer restrictions. This is a de facto privatization of public surveillance, a blurring between private contractors and state entities, and it creates opaque forms of data harvesting that live outside of traditional protective guidelines.[65]

Vigilant has since expanded its "crime-solving" toolkit beyond license plate readers to include ones that claim to recognize faces. In doing so, Vigilant seeks to render human faces as the equivalent of license plates and then feed them back into the policing ecology.[66] Like a network of private detectives, Vigilant creates a God's-eye view of America's interlaced roads and highways, along with everyone who travels along them, while remaining beyond any meaningful form of regulation or accountability.[67]

If we move from the police cruiser to the front porch, we see yet another location where the differences between public and private sector data practices are eroding. A new generation of social media crime-reporting apps like Neighbors, Citizen, and Nextdoor allow users to get alerts about local incidents reported in real time, then discuss them, as well as broadcast, share, and tag security camera footage. Neighbors, which is made by Amazon and relies on its Ring doorbell cameras, defines itself as the "new neighborhood watch" and classifies footage into categories like Crime, Suspicious, or Stranger. Videos are often shared with police.[68] In these residential surveillance ecosystems, the logics of TREASUREMAP and FOXACID conjoin, but connected to the home, the street, and every place in between.

For Amazon, each new Ring device sold helps build yet more large-scale training datasets inside and outside the home, with classificatory logics of normal and anomalous behavior aligned with the battlefield logics of allies and enemies. One example is a feature where users can report stolen Amazon packages. According to one journalistic investigation, many of the posts featured racist commentary, and video posts disproportionately depicted people of color as potential thieves.[69] Beyond reporting crime, Ring is also used to report Amazon employees who are seen as underperforming, such as being insufficiently careful with packages—creating a new layer of worker surveillance and retribution.[70]

To complete its public-private infrastructure of surveillance, Amazon has been aggressively marketing the Ring system to police departments, giving them discounts and offering a portal that allows police to see where Ring cameras are located in the local area and to contact homeowners directly to request footage informally without a warrant.[71] Amazon has negotiated Ring video-sharing partnerships with more than six hundred police departments.[72]

In one case, Amazon negotiated a memorandum of understanding with a police department in Florida, discovered through a public records request filed by journalist Caroline Haskins, which showed that police were incentivized to promote the Neighbors app and for every qualifying download they would receive credits toward free Ring cameras.[73] The result was a "self-perpetuating surveillance network: more people download Neighbors, more people get Ring, surveillance footage proliferates, and police can request whatever they want," Haskins writes.[74] Surveillance capacities that were once ruled over by courts are now on offer in Apple's App Store and promoted by local street cops. As media scholar Tung-Hui Hu observes, by using such apps, we "become freelancers for the state's security apparatus."[75]

Hu describes how targeting—a quintessential militaristic term—in all its forms should be considered together as one interconnected system of power—from targeted advertising to targeting suspicious neighbors to targeting drones. "We cannot merely consider one form of targeting in isolation from the other; conjoined in the sovereignty of data, they call on us to understand power in the age of the cloud differently."[76] The ways of seeing that were once the sole province of intelligence agencies have been granulated and dispersed throughout many social systems—embedded in workplaces, homes, and cars—and promoted by technology companies that live in the cross-hatched spaces that overlap the commercial and military AI sectors.

From Terrorist Credit Scores
to Social Credit Scores

Underlying the military logics of targeting is the idea of the *signature*. Toward the end of President George W. Bush's second term, the CIA argued that it should be able to launch drone attacks based solely on an individual's observed "pattern of behavior" or "signature."[77] Whereas a "personality strike" involves targeting a specific individual, a "signature strike" is when a person is killed due to their metadata signature; in other words, their identity is not known but data suggests that they might be a terrorist.[78] As the Snowden documents showed, during the Obama years, the National Security Agency's global metadata surveillance program would geolocate a SIM card or handset of a suspect, and then the U.S. military would conduct drone strikes to kill the individual in possession of the device.[79] "We kill people based on metadata," said General Michael Hayden, former director of the NSA and the CIA.[80] The NSA's Geo Cell division was reported to use more colorful language: "We track 'em, you whack 'em."[81]

Signature strikes may sound precise and authorized, implying a true mark of someone's identity. But in 2014, the legal organization Reprieve published a report showing that drone strikes attempting to kill 41 individuals resulted in the deaths of an estimated 1,147 people. "Drone strikes have been sold to the American public on the claim that they are 'precise.' But they are only as precise as the intelligence that feeds them," said Jennifer Gibson, who led the report.[82] But the form of the signature strike is not about precision: it is about correlation. Once a pattern is found in the data and it reaches a certain threshold, the suspicion becomes enough to take action even in the absence of definitive proof. This mode of adjudication by pattern recognition is found in many domains — most often taking the form of a score.

Consider an example from the 2015 Syrian refugee crisis. Millions of people were fleeing widespread civil war and enemy occupation in hopes of finding asylum in Europe. Refugees were risking their lives on rafts and overcrowded boats. On September 2, a three-year-old boy named Alan Kurdi drowned in the Mediterranean Sea, alongside his five-year-old brother, when their boat capsized. A photograph showing his body washed up on a beach in Turkey made international headlines as a potent symbol for the extent of the humanitarian crisis: one image standing in for the aggregate horror. But some saw this as a growing threat. It is around this time that IBM was approached about a new project. Could the company use its machine learning platform to detect the data signature of refugees who might be connected to jihadism? In short, could IBM automatically distinguish a terrorist from a refugee?

Andrew Borene, a strategic initiatives executive at IBM, described the rationale behind the program to the military publication *Defense One*: "Our worldwide team, some of the folks in Europe, were getting feedback that there were some concerns that within these asylum-seeking populations that had been starved and dejected, there were fighting-age males coming off of boats that looked awfully healthy. Was that a cause for concern in regard to ISIS and, if so, could this type of solution be helpful?"[83]

From the safe distance of their corporate offices, IBM's data scientists viewed the problem as one best addressed through data extraction and social media analysis. Setting aside the many variables that existed in the conditions of makeshift refugee camps and the dozens of assumptions used to classify terrorist behavior, IBM created an experimental "terrorist credit score" to weed out ISIS fighters from refugees. Analysts harvested a miscellany of unstructured data, from Twitter to the official list of those who had drowned alongside the many capsized boats off the shores of Greece and Turkey. They also made up a data set, modeled on the types of metadata available to border guards. From these disparate measures, they developed a hypothetical threat score: not an absolute indicator of guilt or innocence, they pointed out, but a deep "insight" into the individual, including past addresses, workplaces, and social connections.[84] Meanwhile, Syrian refugees had no knowledge that their personal data was being harvested to trial a system that might single them out as potential terrorists.

This is just one of many cases where new technical systems of state control use the bodies of refugees as test cases. These military and policing logics are now suffused with a form of financialization: socially constructed models of creditworthiness have entered into many AI systems, influencing everything from the ability to get a loan to permission to cross borders. Hundreds of such platforms are now in use around the world, from China to Venezuela to the United States, rewarding predetermined forms of social behavior and penalizing those who do not conform.[85] This "new regime of moralized social classification," in the words of sociologists Marion Fourcade and Kieran Healy, benefits the "high achievers" of the traditional economy while further disadvantaging the least privileged populations.[86] Credit scoring, in the broadest sense, has become a place where the military and commercial signatures combine.

This AI scoring logic is deeply entwined in law enforcement and border control, traditional domains of the state, but it also informs another state function: access to public benefits. As the political scientist Virginia Eubanks shows in her book *Automating Inequality*, when AI systems are deployed as part of the welfare state, they are used primarily as a way to surveil, assess, and restrict people's access to public resources rather than as a way to provide for greater support.[87]

A key example of this dynamic emerged when former Republican governor of Michigan Rick Snyder, previously the chairman of computer hardware computer Gateway, decided to implement two algorithmically driven austerity programs in an attempt to undermine the economic security of his poorest citizens under the auspices of state budget cuts. First, he directed that a matching algorithm be used to implement the state's "fugitive felon" policy, which sought automatically to disqualify individuals from food assistance based on outstanding felony warrants. Between 2012 and 2015, the new system inaccurately matched more than nineteen thousand Michigan residents and automatically disqualified each of them from food assistance.[88]

The second scheme was called the Michigan Integrated Data Automated System (MiDAS), a system built to "robo-adjudicate" and punish those it determined to be defrauding the state's unemployment insurance. MiDAS was designed to treat almost any data discrepancies or inconsistencies in an individual's record as potential evidence of illegal conduct. The system inaccurately identified more than forty thousand Michigan residents of suspected fraud. The consequences were severe: seizure of tax refunds, garnishment of wages, and imposition of civil penalties that were four times the amount people were accused of owing. Ultimately, both systems were giant financial failures, costing Michigan far more money than it saved. Those harmed were able to successfully sue the state over the systems, but not before thousands of people were affected, with many entering bankruptcy.[89]

When viewed in the overall context of state-driven AI systems, one can see the consistent logics between targeting terrorists or undocumented workers and targeting fugitive felons or suspected fraudsters. Even though food assistance and unemployment benefits were created to support the poor and to promote social and economic stability, the use of militaristic systems of command-and-control for the purposes of punishment and exclusion undermine the overall goals of the systems. In essence these systems are punitive, designed on a threat-targeting model. The motifs of scoring and risk have permeated deeply through the structures of state bureaucracy, and the automated decision systems that are imagined in those institutions drive that logic deeply into the way that communities and individuals are imagined, evaluated, scored, and served.

The Tangled Haystack

I am almost at the end of a long day searching through the Snowden archive when I run across a slide that describes the planet as a "haystack of information," in which desirable intel is a needle lost somewhere among the straw. It includes a cheery clip art image of a giant haystack in a field with a blue sky overhead. This cliché of information gathering is tactical: hay is mown for the good of the farm, gleaned to produce value. This invokes a comforting pastoral imagery of data agriculture—tending the fields to further orderly extraction and production cycles. Phil Agre once observed that "technology at present is covert philosophy; the point is to make it openly philosophical."[90] The philosophy here is that data should be extracted globally and structured in order to maintain U.S. hegemony. But we've seen how these stories break down under scrutiny.

The overlapping grids of planetary computation are complex, cross-breeding corporate and state logics, exceeding traditional state border and governance limits, and they are far messier than the idea of winner takes all might imply. As Benjamin Bratton argues, "The armature of planetary-scale computation has a determining logic that is self-reinforcing if not self-fulfilling, and which through the automation of its own infrastructural operations, exceeds any national designs even if it is also used on their behalf."[91] The jingoistic idea of sovereign AI, securely contained within national borders, is a myth. AI infrastructure is already a hybrid, and as Hu argues, so is the labor force underpinning it, from factory laborers in China who make electronic components to Russian programmers providing cloud labor to Moroccan freelancers who screen content and label images.[92]

Taken together, the AI and algorithmic systems used by the state, from the military to the municipal level, reveal a covert philosophy of *en masse* infrastructural command and control via a combination of extractive data techniques, targeting logics, and surveillance. These goals have been central to the intelligence agencies for decades, but now they have spread to many other state functions, from local law enforcement to allocating benefits.[93] This is just part of the deep intermingling of state, municipal, and corporate logics through extractive planetary computation. But it is an uncomfortable bargain: states are making deals with technology companies they can't control or even fully understand, and technology companies are taking on state and extrastate functions that they are ill-suited to fulfill and for which, at some point in the future, they might be held liable.

The Snowden archive shows how far these overlapping and contradictory logics of surveillance extend. One document notes the symptoms of what an NSA employee described as an addiction to the God's-eye view that data seems to offer: "Mountaineers call this phenomenon 'summit fever'—when an 'individual becomes so fixated on reaching the summit that all else fades from consciousness.' I believe that SIGINTers, like the world-class climbers, are not immune to summit fever. It's easy enough to lose sight of the bad weather and push on relentlessly, especially after pouring lots of money, time, and resources into something."[94]

All the money and resources spent on relentless surveillance is part of a fever dream of centralized control that has come at the cost of other visions of social organization. The Snowden disclosures were a watershed moment in revealing how far a culture of extraction can go when the state and the commercial sector collaborate, but the network diagrams and PowerPoint clip art can feel quaint compared to all that has happened since.[95] The NSA's distinctive methods and tools have filtered down to classrooms, police stations, workplaces, and unemployment offices. It is the result of enormous investments, of de facto forms of privatization, and the securitization of risk and fear. The current deep entanglement of different forms of power was the hope of the Third Offset. It has warped far beyond the objective of strategic advantage in battlefield operations to encompass all those parts of everyday life that can be tracked and scored, grounded in normative definitions of how good citizens should communicate, behave, and spend. This shift brings with it a different vision of state sovereignty, modulated by corporate algorithmic governance, and it furthers the profound imbalance of power between agents of the state and the people they are meant to serve.

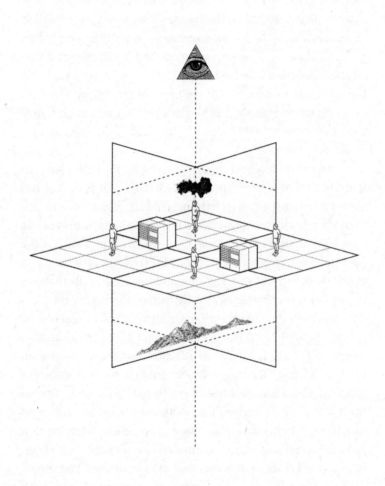

Conclusion
Power

Artificial intelligence is not an objective, universal, or neutral computational technique that makes determinations without human direction. Its systems are embedded in social, political, cultural, and economic worlds, shaped by humans, institutions, and imperatives that determine what they do and how they do it. They are designed to discriminate, to amplify hierarchies, and to encode narrow classifications. When applied in social contexts such as policing, the court system, health care, and education, they can reproduce, optimize, and amplify existing structural inequalities. This is no accident: AI systems are built to see and intervene in the world in ways that primarily benefit the states, institutions, and corporations that they serve. In this sense, AI systems are expressions of power that emerge from wider economic and political forces, created to increase profits and centralize control for those who wield them. But this is not how the story of artificial intelligence is typically told.

The standard accounts of AI often center on a kind of algorithmic exceptionalism—the idea that because AI sys-

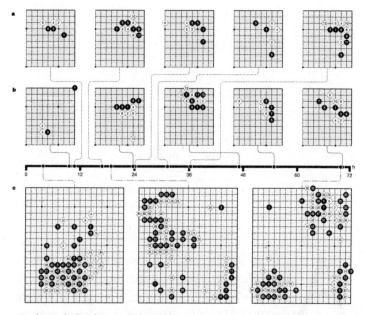

Go knowledge learned by AlphaGo Zero. Courtesy of DeepMind

tems can perform uncanny feats of computation, they must be smarter and more objective than their flawed human creators. Consider this diagram of AlphaGo Zero, an AI program designed by Google's DeepMind to play strategy games.[1] The image shows how it "learned" to play the Chinese strategy game Go by evaluating more than a thousand options per move. In the paper announcing this development, the authors write: "Starting *tabula rasa,* our new program AlphaGo Zero achieved superhuman performance."[2] DeepMind cofounder Demis Hassabis has described these game engines as akin to an alien intelligence. "It doesn't play like a human, but it also doesn't play like computer engines. It plays in a third, almost alien, way. . . . It's like chess from another dimension."[3] When the next iteration mastered Go within three days, Hassabis described it as "rediscovering three thousand years of human knowledge in 72 hours!"[4]

The Go diagram shows no machines, no human workers, no capital investment, no carbon footprint, just an abstract rules-based system endowed with otherworldly skills. Narratives of magic and mystification recur throughout AI's history, drawing bright circles around spectacular displays of speed, efficiency, and computational reasoning.[5] It's no coincidence that one of the iconic examples of contemporary AI is a game.

Games without Frontiers

Games have been a preferred testing ground for AI programs since the 1950s.[6] Unlike everyday life, games offer a closed world with defined parameters and clear victory conditions. The historical roots of AI in World War II stemmed from military-funded research in signal processing and optimization that sought to simplify the world, rendering it more like a strategy game. A strong emphasis on rationalization and prediction emerged, along with a faith that mathematical formalisms would help us understand humans and society.[7] The belief that accurate prediction is fundamentally about reducing the complexity of the world gave rise to an implicit theory of the social: find the signal in the noise and make order from disorder.

This epistemological flattening of complexity into clean signal for the purposes of prediction is now a central logic of machine learning. The historian of technology Alex Campolo and I call this *enchanted determinism*: AI systems are seen as enchanted, beyond the known world, yet deterministic in that they discover patterns that can be applied with predictive certainty to everyday life.[8] In discussions of deep learning systems, where machine learning techniques are extended by layering abstract representations of data on top of each other, enchanted determinism acquires an almost theological quality. That deep learning approaches are often uninterpretable, even to the engineers who created them, gives these systems an aura of being too complex to regulate and too powerful to refuse. As the social anthropologist F. G. Bailey observed, the technique of "obscuring by mystification" is often employed in public settings to argue for a phenomenon's inevitability.[9] We are told to focus on the innovative nature of the method rather than on what is primary: the purpose of the thing itself. Above all, enchanted determinism obscures power and closes off informed public discussion, critical scrutiny, or outright rejection.

Enchanted determinism has two dominant strands, each a mirror image of the other. One is a form of tech utopianism that offers computational interventions as universal solutions applicable to any problem. The other is a tech dystopian perspective that blames algorithms for their negative outcomes as though they are independent agents, without contending with the contexts that shape them and in which they operate. At an extreme, the tech dystopian narrative ends in the singularity, or superintelligence—the theory that a machine intelligence could emerge that will ultimately dominate or destroy humans.[10] This view rarely contends with the reality that so many people around the world are *already* dominated by systems of extractive planetary computation.

These dystopian and utopian discourses are metaphysical twins: one places its faith in AI as a solution to every problem, while the other fears AI as the greatest peril. Each offers a profoundly ahistorical view that locates power solely within technology itself. Whether AI is abstracted as an all-purpose tool or an all-powerful overlord, the result is technological determinism. AI takes the central position in society's redemption or ruin, permitting us to ignore the systemic forces of unfettered neoliberalism, austerity politics, racial inequality, and widespread labor exploitation. Both the tech utopians and dystopians frame the problem with technology always at the center, inevitably expanding into every part of life, decoupled from the forms of power that it magnifies and serves.

When AlphaGo defeats a human grandmaster, it's tempting to imagine that some kind of otherworldly intelligence has arrived. But there's a far simpler and more accurate explanation. AI game engines are designed to play millions of games, run statistical analyses to optimize for winning outcomes, and then play millions more. These programs produce surprising moves uncommon in human games for a straightforward reason: they can play and analyze far more games at a far greater speed than any human can. This is not magic; it is statistical analysis at scale. Yet the tales of preternatural machine intelligence persist.[11] Over and over, we see the ideology of Cartesian dualism in AI: the fantasy that AI systems are disembodied brains that absorb and produce knowledge independently from their creators, infrastructures, and the world at large. These illusions distract from the far more relevant questions: Whom do these systems serve? What are the political economies of their construction? And what are the wider planetary consequences?

The Pipelines of AI

Consider a different illustration of AI: the blueprint for Google's first owned and operated data center, in The Dalles, Oregon. It depicts three 68,680-square-foot buildings, an enormous facility that was estimated in 2008 to use enough energy to power eighty-two thousand homes, or a city the size of Tacoma, Washington.[12] The data center now spreads along the shores of the Columbia River, where it draws heavily on

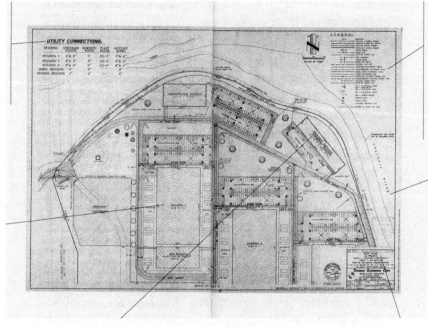

Blueprint of Google Data Center. Courtesy of *Harper's*

some of the cheapest electricity in North America. Google's lobbyists negotiated for six months with local officials to get a deal that included tax exemptions, guarantees of cheap energy, and use of the city-built fiber-optic ring. Unlike the abstract vision of a Go game, the engineering plan reveals how much of Google's technical vision depends on public utilities, including gas mains, sewer pipes, and the high-voltage lines through which the discount electricity would flow. In the words of the writer Ginger Strand, "Through city infrastructure, state give-backs, and federally subsidized power, YouTube is bankrolled by us."[13]

The blueprint reminds us of how much the artificial intelligence industry's expansion has been publicly subsidized:

from defense funding and federal research agencies to public utilities and tax breaks to the data and unpaid labor taken from all who use search engines or post images online. AI began as a major public project of the twentieth century and was relentlessly privatized to produce enormous financial gains for the tiny minority at the top of the extraction pyramid.

These diagrams present two different ways of understanding how AI works. I've argued that there is much at stake in how we define AI, what its boundaries are, and who determines them: it shapes what can be seen and contested. The Go diagram speaks to the industry narratives of an abstract computational cloud, far removed from the earthly resources needed to produce it, a paradigm where technical innovation is lionized, regulation is rejected, and true costs are never revealed. The blueprint points us to the physical infrastructure, but it leaves out the full environmental implications and the political deals that made it possible. These partial accounts of AI represent what philosophers Michael Hardt and Antonio Negri call the "dual operation of *abstraction* and *extraction*" in information capitalism: abstracting away the material conditions of production while extracting more information and resources.[14] The description of AI as fundamentally abstract distances it from the energy, labor, and capital needed to produce it and the many different kinds of mining that enable it.

This book has explored the planetary infrastructure of AI as an extractive industry: from its material genesis to the political economy of its operations to the discourses that support its aura of immateriality and inevitability. We have seen the politics inherent in how AI systems are trained to recognize the world. And we've observed the systemic forms of inequity that make AI what it is today. The core issue is the deep entanglement of technology, capital, and power, of which AI is the latest manifestation. Rather than being inscrutable and alien, these systems are products of larger social and economic structures with profound material consequences.

The Map Is Not the Territory

How do we see the full life cycle of artificial intelligence and the dynamics of power that drive it? We have to go beyond the conventional maps of AI to locate it in a wider landscape. Atlases can provoke a shift in scale, to see how spaces are joined in relation to one another. This book proposes that the real stakes of AI are the global interconnected systems of extraction and power, not the technocratic imaginaries of artificiality, abstraction, and automation. To understand AI for what it is, we need to see the structures of power it serves.

AI is born from salt lakes in Bolivia and mines in Congo, constructed from crowdworker-labeled datasets that seek to classify human actions, emotions, and identities. It is used to navigate drones over Yemen, direct immigration police in the United States, and modulate credit scores of human value and risk across the world. A wide-angle, multiscalar perspective on AI is needed to contend with these overlapping regimes.

This book began below the ground, where the extractive politics of artificial intelligence can be seen at their most literal. Rare earth minerals, water, coal, and oil: the tech sector carves out the earth to fuel its highly energy-intensive infrastructures. AI's carbon footprint is never fully admitted or accounted for by the tech sector, which is simultaneously expanding the networks of data centers while helping the oil and gas industry locate and strip remaining reserves of fossil fuels. The opacity of the larger supply chain for computation in general, and AI in particular, is part of a long-established business model of extracting value from the commons and avoiding restitution for the lasting damage.

Labor represents another form of extraction. In chapter 2, we ventured beyond the highly paid machine learning engineers to consider the other forms of work needed to make artificial intelligence systems function. From the miners extracting tin in Indonesia to crowdworkers in India completing tasks on Amazon Mechanical Turk to iPhone factory workers at Foxconn in China, the labor force of AI is far greater than we normally imagine. Even within the tech companies there is a large shadow workforce of contract laborers, who significantly outnumber full-time employees but have fewer benefits and no job security.[15]

In the logistical nodes of the tech sector, we find humans completing the tasks that machines cannot. Thousands of people are needed to support the illusion of automation: tagging, correcting, evaluating, and editing AI systems to make them appear seamless. Others lift packages, drive for ride-hailing apps, and deliver food. AI systems surveil them all while squeezing the most output from the bare functionality of human bodies: the complex joints of fingers, eyes, and knee sockets are cheaper and easier to acquire than robots. In those spaces, the future of work looks more like the Taylorist factories of the past, but with wristbands that vibrate when workers make errors and penalties given for taking too many bathroom breaks.

The uses of workplace AI further skew power imbalances by placing more control in employers' hands. Apps are used to track workers, nudge them to work longer hours, and rank them in real time. Amazon provides a canonical example of how a microphysics of power—disciplining bodies and their movement through space—is connected to a macrophysics of power, a logistics of planetary time and information. AI systems exploit differences in time and wages across markets to speed the circuits of capital. Suddenly, everyone in urban centers can have—and expects—same day delivery. And the system speeds up again, with the material consequences hidden behind the cardboard boxes, delivery trucks, and "buy now" buttons.

At the data layer, we can see a different geography of extraction. "We are building a mirror of the real world," a Google Street View engineer said in 2012. "Anything that you see in the real world needs to be in our databases."[16] Since then, the harvesting of the real world has only intensified to reach into spaces that were previously hard to capture. As we saw in chapter 3, there has been a widespread pillaging of public spaces; the faces of people in the street have been captured to train facial recognition systems; social media feeds have been ingested to build predictive models of language; sites where people keep personal photos or have online debates have been scraped in order to train machine vision and natural language algorithms. This practice has become so common that few in the AI field even question it. In part, that is because so many careers and market valuations depend on it. The collect-it-all mentality, once the remit of intelligence agencies, is not only normalized but moralized—it is seen as wasteful not to collect data wherever possible.[17]

Once data is extracted and ordered into training sets, it becomes the epistemic foundation by which AI systems classify the world. From the benchmark training sets such as ImageNet, MS-Celeb, or NIST's collections, images are used to represent ideas that are far more relational and contested than the labels may suggest. In chapter 4, we saw how labeling taxonomies allocate people into forced gender binaries, simplistic and offensive racial groupings, and highly normative and stereotypical analyses of character, merit, and emotional state. These classifications, unavoidably value-laden, force a way of seeing onto the world while claiming scientific neutrality.

Datasets in AI are never raw materials to feed algorithms: they are inherently political interventions. The entire practice of harvesting data, categorizing and labeling it, and then using it to train systems is a form of politics. It has brought a shift to what are called operational images—representations of the world made solely for machines.[18] Bias is a symptom of a deeper affliction: a far-ranging and centralizing normative logic that is used to determine how the world should be seen and evaluated.

A central example of this is affect detection, described in chapter 5, which draws on controversial ideas about the relation of faces to emotions and applies them with the reductive logic of a lie detector test. The science remains deeply contested.[19] Institutions have always classified people into identity categories, narrowing personhood and cutting it down into precisely measured boxes. Machine learning allows that to happen at scale. From the hill towns of Papua New Guinea to military labs in Maryland, techniques have been developed to reduce the messiness of feelings, interior states, preferences, and identifications into something quantitative, detectable, and trackable.

What epistemological violence is necessary to make the world readable to a machine learning system? AI seeks to systematize the unsystematizable, formalize the social, and convert an infinitely complex and changing universe into a Linnaean order of machine-readable tables. Many of AI's achievements have depended on boiling things down to a terse set of formalisms based on proxies: identifying and naming some features while ignoring or obscuring countless others. To adapt a phrase from philosopher Babette Babich, machine learning exploits what it does know to predict what it does not know: a game of repeated approximations. Datasets are also *proxies*—stand-ins for what they claim to measure. Put simply, this is transmuting difference into computable sameness. This kind of knowledge schema recalls what Friedrich Nietzsche described as "the falsifying of the multifarious and incalculable into the identical, similar, and calculable."[20] AI systems become deterministic when these proxies are taken as ground truth, when fixed labels are applied to a fluid complexity. We saw this in the cases where AI is used to predict gender, race, or sexuality from a photograph of a face.[21] These approaches resemble phrenology and physiognomy in their desire to essentialize and impose identities based on external appearances.

The problem of ground truth for AI systems is heightened in the context of state power, as we saw in chapter 6. The intelligence agencies led the way on the mass collection of data, where metadata signatures are sufficient for lethal drone strikes and a cell phone location becomes a proxy for an unknown target. Even here, the bloodless language of metadata and surgical strikes is directly contradicted by the unintended killings from drone missiles.[22] As Lucy Suchman has asked, how are "objects" identified as imminent threats? We know that "ISIS pickup truck" is a category based on hand-labeled data, but who chose the categories and identified the vehicles?[23] We saw the epistemological confusions and errors of object recognition training sets like ImageNet; military AI systems and drone attacks are built on the same unstable terrain.

The deep interconnections between the tech sector and the military are now framed within a strong nationalist agenda. The rhetoric about the AI war between the United States and China drives the interests of the largest tech companies to operate with greater government support and few restrictions. Meanwhile, the surveillance armory used by agencies like the NSA and the CIA is now deployed domestically at a municipal level in the in-between space of commercial-military contracting by companies like Palantir. Undocumented immigrants are hunted down with logistical systems of total information control and capture that were once reserved for extralegal espionage. Welfare decision-making systems are used to track anomalous data patterns in order to cut people off from unemployment benefits and accuse them of fraud. License plate reader technology is being used by home surveillance systems—a widespread integration of previously separate surveillance networks.[24]

The result is a profound and rapid expansion of surveillance and a blurring between private contractors, law enforcement, and the tech sector, fueled by kickbacks and secret deals. It is a radical redrawing of civic life, where the centers of power are strengthened by tools that see with the logics of capital, policing, and militarization.

Toward Connected Movements for Justice

If AI currently serves the existing structures of power, an obvious question might be: Should we not seek to democratize it? Could there not be an AI for the people that is reoriented toward justice and equality rather than industrial extraction and discrimination? This may seem appealing, but as we have seen throughout this book, the infrastructures and forms of power that enable and are enabled by AI skew strongly toward the centralization of control. To suggest that we democratize AI to reduce asymmetries of power is a little like arguing for democratizing weapons manufacturing in the service of peace. As Audre Lorde reminds us, the master's tools will never dismantle the master's house.[25]

A reckoning is due for the technology sector. To date, one common industry response has been to sign AI ethics principles. As European Union parliamentarian Marietje Schaake observed, in 2019 there were 128 frameworks for AI ethics in Europe alone.[26] These documents are often presented as products of a "wider consensus" on AI ethics. But they are overwhelmingly produced by economically developed countries, with little representation from Africa, South and Central America, or Central Asia. The voices of the people most harmed by AI systems are largely missing from the processes that produce them.[27] Further, ethical principles and statements don't discuss how they should be implemented, and they are rarely enforceable or accountable to a broader public. As Shannon Mattern has noted, the focus is more commonly on the ethical ends for AI, without assessing the ethical means of its application.[28] Unlike medicine or law, AI has no formal professional governance structure or norms—no agreed-upon definitions and goals for the field or standard protocols for enforcing ethical practice.[29]

Self-regulating ethical frameworks allow companies to choose how to deploy technologies and, by extension, to decide what ethical AI means for the rest of the world.[30] Tech companies rarely suffer serious financial penalties when their AI systems violate the law and even fewer consequences when their ethical principles are violated. Further, public companies are pressured by shareholders to maximize return on investment over ethical concerns, commonly making ethics secondary to profits. As a result, ethics is necessary but not sufficient to address the fundamental concerns raised in this book.

To understand what is at stake, we must focus less on ethics and more on power. AI is invariably designed to amplify and reproduce the forms of power it has been deployed to optimize. Countering that requires centering the interests of the communities most affected.[31] Instead of glorifying company founders, venture capitalists, and technical visionaries, we should begin with the lived experiences of those who are disempowered, discriminated against, and harmed by AI systems. When someone says, "AI ethics," we should assess the labor conditions for miners, contractors, and crowdworkers. When we hear "optimization," we should ask if these are tools for the inhumane treatment of immigrants. When there is applause for "large-scale automation," we should remember the resulting carbon footprint at a time when the planet is already under extreme stress. What would it mean to work toward justice across all these systems?

In 1986, the political theorist Langdon Winner described a society "committed to making artificial realities" with no concern for the harms it could bring to the conditions of life: "Vast transformations in the structure of our common world have been undertaken with little attention to what those alterations mean. . . . In the technical realm we repeatedly enter into a series of social contracts, the terms of which are only revealed after signing."[32]

In the four decades since, those transformations are now at a scale that has shifted the chemical composition of the atmosphere, the temperature of Earth's surface, and the contents of the planet's crust. The gap between how technology is judged on its release and its lasting consequences has only widened. The social contract, to the extent that there ever was one, has brought a climate crisis, soaring wealth inequality, racial discrimination, and widespread surveillance and labor exploitation. But the idea that these transformations occurred in ignorance of their possible results is part of the problem. The philosopher Achille Mbembé sharply critiques the idea that we could not have foreseen what would become of the knowledge systems of the twenty-first century, as they were always "operations of abstraction that claim to rationalize the world on the basis of corporate logic."[33] He writes: "It is about extraction, capture, the cult of data, the commodification of human capacity for thought and the dismissal of critical reason in favour of programming. . . . Now more than ever before, what we need is a new critique of technology, of the experience of technical life."[34]

The next era of critique will also need to find spaces beyond technical life by overturning the dogma of inevitability. When AI's rapid expansion is seen as unstoppable, it is possible only to patch together legal and technical restraints on systems after the fact: to clean up datasets, strengthen privacy laws, or create ethics boards. But these will always be partial and incomplete responses in which technology is assumed and everything else must adapt. But what happens if we reverse this polarity and begin with the commitment to a more just and sustainable world? How can we intervene to address interdependent issues of social, economic, and climate injustice? Where does technology serve that vision? And are there places where AI should not be used, where it undermines justice?

This is the basis for a renewed politics of refusal—opposing the narratives of technological inevitability that says, "If it can be done, it will be." Rather than asking where AI will be applied, merely because it can, the emphasis should be on *why* it ought to be applied. By asking, "Why use artificial intelligence?" we can question the idea that everything should be subject to the logics of statistical prediction and profit accumulation, what Donna Haraway terms the "informatics of domination."[35] We see glimpses of this refusal when populations choose to dismantle predictive policing, ban facial recognition, or protest algorithmic grading. So far these minor victories have been piecemeal and localized, often centered in cities with more resources to organize, such as London, San Francisco, Hong Kong, and Portland, Oregon. But they point to the need for broader national and international movements that refuse technology-first approaches and focus on addressing underlying inequities and injustices. Refusal requires rejecting the idea that the same tools that serve capital, militaries, and police are also fit to transform schools, hospitals, cities, and ecologies, as though they were value neutral calculators that can be applied everywhere.

The calls for labor, climate, and data justice are at their most powerful when they are united. Above all, I see the greatest hope in the growing justice movements that address the interrelatedness of capitalism, computation, and control: bringing together issues of climate justice, labor rights, racial justice, data protection, and the overreach of police and military power. By rejecting systems that further inequity and violence, we challenge the structures of power that AI currently reinforces and create the foundations for a different society.[36] As Ruha Benjamin notes, "Derrick Bell said it like this: 'To see things as they really are, you must imagine them for what they might be.' We are pattern makers and we must change the content of our existing patterns."[37] To do so will require shaking off the enchantments of tech solutionism and embracing alternative solidarities—what Mbembé calls "a different politics of inhabiting the Earth, of repairing and sharing the planet."[38] There are sustainable collective politics beyond value extraction; there are commons worth keeping, worlds beyond the market, and ways to live beyond discrimination and brutal modes of optimization. Our task is to chart a course there.

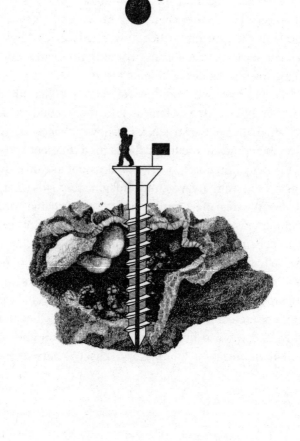

Coda
Space

Acountdown begins. File footage starts rolling. Engines at the base of a towering Saturn V ignite, and the rocket begins liftoff. We hear the voice of Jeff Bezos: "Ever since I was five years old—that's when Neil Armstrong stepped onto the surface of the moon—I've been passionate about space, rockets, rocket engines, space travel." A parade of inspirational images appears: mountain climbers at summits, explorers descending into canyons, an ocean diver swimming through a shoal of fish.

Cut to Bezos in a control room during a launch, adjusting his headset. His voiceover continues: "This is the most important work I'm doing. It's a simple argument, this is the best planet. And so we face a choice. As we move forward, we're gonna have to decide whether we want a civilization of stasis—we will have to cap population, we will have to cap energy usage per capita—or we can fix that problem, by moving out into space."[1]

The soundtrack soars, and images of deep space are counterposed with shots of the busy freeways of Los Angeles and clogged cloverleaf junctions. "Von Braun said, after the lunar landing, 'I have learned to use the word impossible with great caution.' And I hope you guys take that attitude about your lives."[2]

This scene comes from a promotional video for Bezos's private aerospace company, Blue Origin. The company motto is *Gradatim Ferociter,* Latin for "Step by Step, Ferociously." In the near term, Blue Origin is building reusable rockets and lunar landers, testing them primarily at its facility and sub-orbital base in West Texas. By 2024, the company wants to be shuttling astronauts and cargo to the Moon.[3] But in the longer term, the company's mission is far more ambitious: to help bring about a future in which millions are living and working in space. Specifically, Bezos has outlined his hopes to build giant space colonies, where people would live in floating manufactured environments.[4] Heavy industry would move off-planet altogether, the new frontier for extraction. Meanwhile, Earth would be zoned for residential building and light industry, left as a "beautiful place to live, a beautiful place to visit"—presumably for those who can afford to be there, rather than working in the off-world colonies.[5]

Bezos possesses extraordinary and growing industrial power. Amazon continues to capture more of U.S. online commerce, Amazon Web Services represents nearly half of the cloud-computing industry, and, by some estimates, Amazon's site has more product searches than Google.[6] Despite all this, Bezos is worried. His fear is that the planet's growing energy demands will soon outstrip its limited supply. For him, the greatest concern "is not necessarily extinction" but *stasis:* "We will have to stop growing, which I think is a very bad future."[7]

Bezos is not alone. He is just one of several tech billionaires focused on space. Planetary Resources, led by the founder of the X Prize, Peter Diamandis, and backed with investment from Google's Larry Page and Eric Schmidt, aimed to create the first commercial mine in space by drilling asteroids.[8] Elon Musk, chief executive of Tesla and SpaceX, has announced his intention to colonize Mars within a hundred years—while admitting that, to do so, the first astronauts must "be prepared to die."[9] Musk has also advocated terraforming the surface of Mars for human settlement by exploding nuclear weapons at the poles.[10] SpaceX made a T-shirt that reads "NUKE MARS." Musk also conducted what is arguably the most expensive public relations exercise in history when he launched a Tesla car into heliocentric orbit on a SpaceX Falcon Heavy rocket. Researchers estimate that the car will remain in space for millions of years, until it finally crashes back to Earth.[11]

The ideology of these space spectacles is deeply interconnected with that of the AI industry. Extreme wealth and power generated from technology companies now enables a small group of men to pursue their own private space race. They depend on exploiting the knowledge and infrastructures of the public space programs of the twentieth century and often rely on government funding and tax incentives as well.[12] Their aim is not to limit extraction and growth but to extend it across the solar system. In actuality, these efforts are as much about an *imaginary* of space, endless growth, and immortality than they are about the uncertain and unpleasant possibilities of actual space colonization.

Bezos's inspiration for conquering space comes, in part, from the physicist and science fiction novelist Gerard K. O'Neill. O'Neill wrote *The High Frontier: Human Colonies in Space*, a 1976 fantasy of space colonization, which includes lush illustrations of moon mining with Rockwellian abundance.[13] Bezos's plan for Blue Origin is inspired by this bucolic vision

of permanent human settlement, for which no current technology exists.[14] O'Neill was driven by the "dismay and shock"
he felt when he read the 1972 landmark report by the Club
of Rome, called *The Limits to Growth*.[15] The report published
extensive data and predictive models about the end of nonrenewable resources and the impact on population growth,
sustainability, and humanity's future on Earth.[16] As the architecture and planning scholar Fred Scharmen summarizes:

> The Club of Rome models calculate outcomes from
> different sets of initial assumptions. The baseline
> scenarios, extrapolated from then-current trends,
> show resource and population collapse before the
> year 2100. When the models assume double the
> known resource reserves, they collapse again, to a
> slightly higher level but still before 2100. When they
> assume that technology will make available "un-
> limited" resources, population collapses *even more
> sharply* than before due to spikes in pollution. With
> pollution controls added to the model, population
> collapses after running out of food. In models that
> increase agricultural capacity, pollution overruns
> previous controls and both food and population
> collapse.[17]

Limits to Growth suggested that moving to sustainable
management and reuse of resources was the answer to long-
term stability of global society and that narrowing the gap between rich and poor nations was the key to survival. Where
Limits to Growth fell short was that it did not foresee the larger
set of interconnected systems that now make up the global
economy and how previously uneconomic forms of mining
would be incentivized, driving greater environmental harms,

land and water degradation, and accelerated resource depletion.

In writing *The High Frontier,* O'Neill wanted to imagine a different way out of the no-growth model rather than
limiting production and consumption.[18] By positing that space
was a solution, O'Neill redirected global anxiety in the 1970s
over gasoline shortages and oil crises with visions of serene
stable space structures that would simultaneously preserve the
status quo and offer new opportunities. "If Earth doesn't have
enough surface area," O'Neill urged, "then humans should
simply build more."[19] The science of how it would work and
the economics of how we could afford it were details left for
another day; the dream was all that mattered.[20]

That space colonization and frontier mining have become the common corporate fantasies of tech billionaires
underscores a fundamentally troubling relationship to Earth.
Their vision of the future does not include minimizing oil and
gas exploration or containing resource consumption or even
reducing the exploitative labor practices that have enriched
them. Instead, the language of the tech elite often echoes settler
colonialism, seeking to displace Earth's population and capture territory for mineral extraction. Silicon Valley's billionaire
space race similarly assumes that the last commons—outer
space—can be taken by whichever empire gets there first. This
is despite the main convention governing space mining, the
1967 Outer Space Treaty, which recognizes that space is the
"common interest of all mankind" and that any exploration or
use "should be carried on for the benefit of all peoples."[21]

In 2015, Bezos's Blue Origin and Musk's SpaceX lobbied
Congress and the Obama administration to enact the Commercial Space Launch Competitiveness Act.[22] It extends an ex-
emption for commercial space companies from federal regulation until 2023, allowing them to own any mining resources

extracted from asteroids and keep the profits.[23] This legislation directly undercuts the idea of space as a commons, and creates a commercial incentive to "go forth and conquer."[24]

Space has become the ultimate imperial ambition, symbolizing an escape from the limits of Earth, bodies, and regulation. It is perhaps no surprise that many of the Silicon Valley tech elite are invested in the vision of abandoning the planet. Space colonization fits well alongside the other fantasies of life-extension dieting, blood transfusions from teenagers, brain-uploading to the cloud, and vitamins for immortality.[25] Blue Origin's high-gloss advertising is part of this dark utopianism. It is a whispered summons to become the Übermensch, to exceed all boundaries: biological, social, ethical, and ecological. But underneath, these visions of brave new worlds seem driven most of all by fear: fear of death—individually and collectively—and fear that time is truly running out.

I'm back in the van for the last leg of my journey. I drive south out of Albuquerque, New Mexico, headed toward the Texas border. On my way, I take a detour past the rocky face of San Augustin Peak and follow the steep drive down to the White Sands Missile Range, where in 1946 the United States launched the first rocket containing a camera into space. That mission was led by Wernher von Braun, who had been the technical director of Germany's missile rocket development program. He defected to the United States after the war, and there he began experimenting with confiscated V-2 rockets—the very missiles he had helped design, which had been fired against the Allies across Europe. But this time he sent them directly upward, into space. The rocket ascended to an altitude of 65 miles, capturing images every 1.5 seconds, before crashing into the New Mexican desert. The film survived inside of a steel cassette, revealing a grainy but distinctly Earthlike curve.[26]

View of Earth from a camera on V-2 #13,
launched October 24, 1946. Courtesy White Sands
Missile Range/Applied Physics Laboratory

That Bezos chose to quote von Braun in his Blue Origin
commercial is notable. Von Braun was chief rocket engineer
of the Third Reich and admitted using concentration camp
slave labor to build his V-2 rockets; some consider him a war
criminal.[27] More people died in the camps building the rockets
than were killed by them in war.[28] But it is von Braun's work
as head of NASA's Marshall Space Flight Center, where he was
instrumental in the design of the Saturn V rocket, that is best
known.[29] Bathed in the glow of Apollo 11, washed clean of history, Bezos has found his hero—a man who refused to believe
in impossibility.

After driving through El Paso, Texas, I take Route 62
toward the Salt Basin Dunes. It's late in the afternoon, and
colors are starting to bloom in the cumulus clouds. There's a
T-junction, and after turning right, the road begins to trace
along the Sierra Diablos. This is Bezos country. The first in-

Blue Origin suborbital launch facility, West Texas.
Photograph by Kate Crawford

dication is a large ranch house set back from the road, with a
sign in red letters that reads "Figure 2" on a white gate. It's the
ranch that Bezos purchased in 2004, just part of the three hundred thousand acres he owns in the area.[30] The land has a violent colonial history: one of the final battles between the Texas
Rangers and the Apaches occurred just west of this site in 1881,
and nine years later the ranch was created by the one-time
Confederate rider and cattleman James Monroe Daugherty.[31]

Nearby is the turn-off to the Blue Origin suborbital
launch facility. The private road is blocked by a bright blue gate
with security notices warning of video surveillance and a guard
station bristling with cameras. I stay on the highway and pull
the van over to the side of the road a few minutes away. From
here, the views stretch across the valley to the Blue Origin land-

ing site, where the rockets are being tested for what is expected to be the company's first human mission into space. Cars pass through the boom gates as the workers clock out for the day.

Looking back at the clusters of sheds that mark out the rocket base, it feels very provisional and makeshift in this dry expanse of the Permian Basin. The vast span of the valley is broken with a hollow circle, the landing pad where Blue Origin's reusable rockets are meant to touch down on a feather logo painted in the center. That's all there is to see. It's a private infrastructure-in-progress, guarded and gated, a technoscientific imaginary of power, extraction, and escape, driven by the wealthiest man on the planet. It is a hedge against Earth.

The light is fading now, and steel-gray clouds are moving against the sky. The desert looks silvery, dotted with white sage bushes and clusters of volcanic tuff punctuating what was once the floor of a great inland sea. After taking a photograph, I head back to the van to begin the final drive of the day to the town of Marfa. It's not until I start driving away that I realize I'm being followed. Two matching black Chevrolet pickups begin aggressively tailgating at close range. I pull over in the hope they will pass. They also pull over. No one moves. After waiting a few minutes, I slowly begin to drive again. They maintain their sinister escort all the way to the edge of the darkening valley.

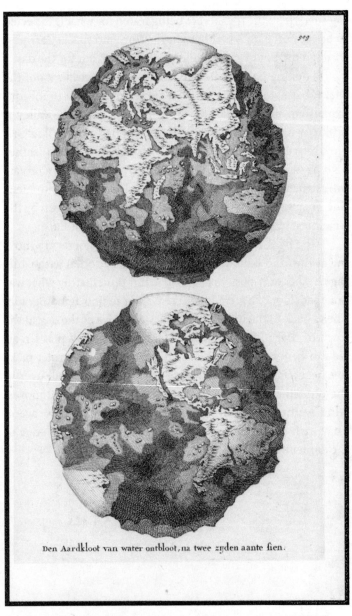

Den Aardkloot van water ontbloot, na twee zyden aante fien.

The World without Water (Den Aardkloot van water ontbloot),
Thomas Burnet's 1694 map of the world drained of its oceans.

Acknowledgments

All books are collective projects, and the longer they take to write, the bigger the collective. *Atlas of AI* was many years in the making, and was made possible thanks to the friends, colleagues, collaborators, and coadventurers who came along with me. There were many late-night conversations and early-morning coffees, as well as road trips and roundtables, all of which have brought this book to life. I have enough gratitude to merit a separate volume, but these few words will have to suffice for now.

First, to the scholars and friends whose work left the deepest imprints on this book: Mike Ananny, Geoffrey Bowker, Benjamin Bratton, Simone Browne, Wendy Chun, Vladan Joler, Alondra Nelson, Jonathan Sterne, Lucy Suchman, Fred Turner, and McKenzie Wark. To Jer Thorp, thank you for the days we wrote side by side, and for the commiserations and celebrations (depending on the week).

I've been fortunate over the years to be a member of multiple research communities who have taught me so much. There are many scholars and engineers who make Microsoft Research an exceptional place, and I'm grateful to be a member of both the FATE group and the Social Media Collective. Thanks to Ifeoma Ajunwa, Peter Bailey, Solon Barocas, Nancy

Baym, Christian Borgs, Margarita Boyarskaya, danah boyd, Sarah Brayne, Jed Brubaker, Bill Buxton, Jennifer Chayes, Tressie McMillan Cottom, Hal Daume, Jade Davis, Fernando Diaz, Kevin Driscoll, Miro Dudik, Susan Dumais, Megan Finn, Timnit Gebru, Tarleton Gillespie, Mary L. Gray, Dan Greene, Caroline Jack, Adam Kalai, Tero Karppi, Os Keyes, Airi Lampinen, Jessa Lingel, Sonia Livingstone, Michael Madaio, Alice Marwick, J. Nathan Matias, Josh McVeigh-Schultz, Andrés Monroy-Hernández, Dylan Mulvin, Laura Norén, Alexandra Olteanu, Aaron Plasek, Nick Seaver, Aaron Shapiro, Luke Stark, Lana Swartz, TL Taylor, Jenn Wortman Vaughan, Hanna Wallach, and Glen Weyl. I could not ask to learn from a more luminous constellation of scholars.

Particular thanks to everyone who has been a part of creating the AI Now Institute at NYU: Alejandro Calcaño Bertorelli, Alex Butzbach, Roel Dobbe, Theodora Dryer, Genevieve Fried, Casey Gollan, Ben Green, Joan Greenbaum, Amba Kak, Elizabeth Kaziunas, Varoon Mathur, Erin McElroy, Andrea Nill Sánchez, Mariah Peebles, Deb Raji, Joy Lisi Rankin, Noopur Raval, Dillon Reisman, Rashida Richardson, Julia Bloch Thibaud, Nantina Vgontzas, Sarah Myers West, and Meredith Whittaker.

And I'm always grateful to the extraordinary Australian scholars who grounded me from the beginning, including Kath Albury, Mark Andrejevic, Genevieve Bell, Jean Burgess, Chris Chesher, Anne Dunn, Gerard Goggin, Melissa Gregg, Larissa Hjorth, Catharine Lumby, Elspeth Probyn, Jo Tacchi, and Graeme Turner. The road is long, but it always leads back home.

This book greatly benefited from several research assistants, readers, and archivists over the years, all of whom are amazing scholars in their own right. Thanks to Sally Collings, Sarah Hamid, Rebecca Hoffman, Caren Litherland, Kate Miltner, Léa Saint-Raymond, and Kiran Samuel for helping me think harder, track down sources, access archives, and complete endnotes. Particular thanks to Alex Campolo for his depth of expertise on the history of science in the twentieth century—it is a joy to work with you. Elmo Keep was a brilliant interlocutor, and Joy Lisi Rankin was an insightful editor. Several archivists generously helped this project, but particularly Janet Monge at the Samuel Morton skull archive and Henrik Moltke with the Snowden archive.

To Joseph Calamia, I owe you so much. Thank you for believing in this project and for your patience while I completed the many journeys it required. Thanks also to Bill Frucht and Karen Olson at Yale University Press for bringing it over the line.

I'm deeply indebted to the institutions that invited me to visit and gave me time to write. Thanks to the École Normale Supérieure in Paris, where I was the inaugural chair in AI and Justice, to the Robert Bosch Academy in Berlin, where I was a Richard von Weizsäcker Fellow, and to the University of Melbourne for the Miengunyah Distinguished Visiting Fellowship. The communities at each one of these institutions have been so welcoming and expanded the contexts of this atlas. For making all that possible, thanks to Anne Bouverot, Tanya Perelmuter, Mark Mezard, Fondation Abeona, Sandra Breka, Jannik Rust, and Jeannie Paterson.

I developed the ideas in this book in conference presentations, exhibitions, and lectures over a decade, across the fields of architecture, art, critical geography, computer science, cultural studies, law, media studies, philosophy, and science and technology studies. Audiences at the Australian National University, California Institute of Technology, Columbia University, Haus der Kulturen der Welt, MIT, National Academy of Science, New York University, Royal Society of London,

Smithsonian Museum, University of New South Wales, Yale University, École Normale Supérieure, and at conferences like NeurIPS, AoIR, and ICML gave vital feedback as I developed this project.

Some material in various chapters has been drawn from previously published journal articles and substantially altered for this context, and I'd like to acknowledge all the coauthors and journals with whom it's been my honor to collaborate:

"Enchanted Determinism: Power without Responsibility in Artificial Intelligence," *Engaging Science, Technology, and Society* 6 (2020): 1–19 (with Alex Campolo); "'Excavating AI: The Politics of Images in Machine Learning Training Sets," *AI and Society* 2020 (with Trevor Paglen); "Alexa, Tell Me about Your Mother: The History of the Secretary and the End of Secrecy," *Catalyst: Feminism, Theory, Technoscience* 6, no. 1 (2020) (with Jessa Lingel); "AI Systems as State Actors," *Columbia Law Review* 119 (2019): 1941–72 (with Jason Schultz); "Halt the Use of Facial-Recognition Technology until It Is Regulated," *Nature* 572 (2019): 565; "Dirty Data, Bad Predictions: How Civil Rights Violations Impact Police Data, Predictive Policing Systems, and Justice," *NYU Law Review Online* 94, no. 15 (2019): 15–55 (with Rashida Richardson and Jason Schultz); "Anatomy of an AI System: The Amazon Echo as an Anatomical Map of Human Labor, Data and Planetary Resources," *AI Now Institute* and *Share Lab*, September 7, 2018 (with Vladan Joler); "Datasheets for Datasets," Proceedings of the Fifth Workshop on Fairness, Accountability, and Transparency in Machine Learning, Stockholm, 2018 (with Timnit Gebru, Jamie Morgenstern, Briana Vecchione, Jennifer Wortman Vaughan, Hanna Wallach, and Hal Daumeé III); "The Problem with Bias: Allocative Versus Representational Harms in Machine Learning," SIGCIS Conference 2017 (with Solon Barocas, Aaron Shapiro, and Hanna Wallach); "Limitless Worker Surveillance,"

California Law Review 105, no. 3 (2017): 735–76 (with Ifeoma Ajunwa and Jason Schultz); "Can an Algorithm Be Agonistic? Ten Scenes from Life in Calculated Publics," *Science, Technology and Human Values* 41 (2016): 77–92; "Asking the Oracle," in *Astro Noise,* ed. Laura Poitras (New Haven: Yale University Press, 2016), 128–41; "Seeing without Knowing: Limitations of the Transparency Ideal and Its Application to Algorithmic Accountability," *New Media and Society* 20, no. 3 (2018): 973–89 (with Mike Ananny); "Where Are the Human Subjects in Big Data Research? The Emerging Ethics Divide," *Big Data and Society* 3, no. 1 (2016) (with Jake Metcalf); "Exploring or Exploiting? Social and Ethical Implications of Autonomous Experimentation in AI," Workshop on Fairness, Accountability, and Transparency in Machine Learning (FAccT), 2016 (with Sarah Bird, Solon Barocas, Fernando Diaz, and Hanna Wallach); "There Is a Blind Spot in AI Research," *Nature* 538 (2016): 311–13 (with Ryan Calo); "Circuits of Labour: A Labour Theory of the iPhone Era," *TripleC: Communication, Capitalism and Critique,* 2014 (with Jack Qiu and Melissa Gregg); "Big Data and Due Process: Toward a Framework to Redress Predictive Privacy Harms," *Boston College Law Review* 55, no. 1 (2014) (with Jason Schultz); and "Critiquing Big Data: Politics, Ethics, Epistemology," *International Journal of Communications* 8 (2014): 663–72 (with Kate Miltner and Mary Gray).

Beyond articles, I've been fortunate to participate on collaborative reports with the team at the AI Now Institute, which have informed this book: *AI Now 2019 Report,* AI Now Institute, 2019 (with Roel Dobbe, Theodora Dryer, Genevieve Fried, Ben Green, Amba Kak, Elizabeth Kaziunas, Varoon Mathur, Erin McElroy, Andrea Nill Sánchez, Deborah Raji, Joy Lisi Rankin, Rashida Richardson, Jason Schultz, Sarah Myers West, and Meredith Whittaker); "Discriminating Systems: Gender, Race and Power in AI," AI Now Institute, 2019 (with Sarah

Myers West and Meredith Whittaker); *AI Now Report 2018*, AI Now Institute, 2018 (with Meredith Whittaker, Roel Dobbe, Genevieve Fried, Elizabeth Kaziunas, Varoon Mathur, Sarah Myers West, Rashida Richardson, Jason Schultz, and Oscar Schwartz); "Algorithmic Impact Assessments: A Practical Framework for Public Agency Accountability," AI Now Institute, 2018 (with Dillon Reisman, Jason Schultz, and Meredith Whittaker); *AI Now 2017 Report*, AI Now Institute, 2017 (with Alex Campolo, Madelyn Sanfilippo, and Meredith Whittaker); and *AI Now 2016 Report*, NYU Information Law Institute, 2016 (with Madeleine Clare Elish, Solon Barocas, Aaron Plasek, Kadija Ferryman, and Meredith Whittaker).

Finally, this book would not exist without these people: Trevor Paglen, a true compass, from desert explorations to archaeological investigations; Vladan Joler, a friend in map-making, whose designs illuminate this book and my thinking; Laura Poitras, who gave me the courage; Karen Murphy, for her designer's eye; Adrian Hobbes and Edwina Throsby, for getting me through the fires; Bo Daley, who made everything better; and to my family, Margaret, James, Judith, Claudia, Cliff, and Hilary. Eternal thanks are due to Jason and Elliott, my favorite cartographers.

Notes

Introduction

1. Heyn, "Berlin's Wonderful Horse."
2. Pfungst, *Clever Hans.*
3. "'Clever Hans' Again."
4. Pfungst, *Clever Hans.*
5. Pfungst.
6. Lapuschkin et al., "Unmasking Clever Hans Predictors."
7. See the work of philosopher Val Plumwood on the dualisms of intelligence-stupid, emotional-rational, and master-slave. Plumwood, "Politics of Reason."
8. Turing, "Computing Machinery and Intelligence."
9. Von Neumann, *The Computer and the Brain,* 44. This approach was deeply critiqued by Dreyfus, *What Computers Can't Do.*
10. See Weizenbaum, "On the Impact of the Computer on Society," 612. After his death, Minsky was implicated in serious allegations related to convicted pedophile and rapist Jeffrey Epstein. Minsky was one of several scientists who met with Epstein and visited his island retreat where underage girls were forced to have sex with members of Epstein's coterie. As scholar Meredith Broussard observes, this was part of a broader culture of exclusion that became endemic in AI: "As wonderfully creative as Minsky and his cohort were, they also solidified the culture of tech as a billionaire boys' club. Math, physics, and the other 'hard' sciences have never been hospitable to women and people of color; tech followed this lead." See Broussard, *Artificial Unintelligence,* 174.
11. Weizenbaum, *Computer Power and Human Reason,* 202–3.
12. Greenberger, *Management and the Computer of the Future,* 315.
13. Dreyfus, *Alchemy and Artificial Intelligence.*

14. Dreyfus, *What Computers Can't Do.*

15. Ullman, *Life in Code,* 136–37.

16. See, as one of many examples, Poggio et al., "Why and When Can Deep—but Not Shallow—Networks Avoid the Curse of Dimensionality."

17. Quoted in Gill, *Artificial Intelligence for Society,* 3.

18. Russell and Norvig, *Artificial Intelligence,* 30.

19. Daston, "Cloud Physiognomy."

20. Didi-Huberman, *Atlas,* 5.

21. Didi-Huberman, 11.

22. Franklin and Swenarchuk, *Ursula Franklin Reader,* Prelude.

23. For an account of the practices of data colonization, see "Colonized by Data"; and Mbembé, *Critique of Black Reason.*

24. Fei-Fei Li quoted in Gershgorn, "Data That Transformed AI Research."

25. Russell and Norvig, *Artificial Intelligence,* 1.

26. Bledsoe quoted in McCorduck, *Machines Who Think,* 136.

27. Mattern, *Code and Clay, Data and Dirt,* xxxiv–xxxv.

28. Ananny and Crawford, "Seeing without Knowing."

29. Any list will always be an inadequate account of all the people and communities who have inspired and informed this work. I'm particularly grateful to these research communities: FATE (Fairness, Accountability, Transparency and Ethics) and the Social Media Collective at Microsoft Research, the AI Now Institute at NYU, the Foundations of AI working group at the École Normale Supérieure, and the Richard von Weizsäcker Visiting Fellows at the Robert Bosch Academy in Berlin.

30. Saville, "Towards Humble Geographies."

31. For more on crowdworkers, see Gray and Suri, *Ghost Work;* and Roberts, *Behind the Screen.*

32. Canales, *Tenth of a Second.*

33. Zuboff, *Age of Surveillance Capitalism.*

34. Cetina, *Epistemic Cultures,* 3.

35. "Emotion Detection and Recognition (EDR) Market Size."

36. Nelson, Tu, and Hines, "Introduction," 5.

37. Danowski and de Castro, *Ends of the World.*

38. Franklin, *Real World of Technology,* 5.

1

Earth

1. Brechin, *Imperial San Francisco.*

2. Brechin, 29.

3. Agricola quoted in Brechin, 25.

4. Quoted in Brechin, 50.

5. Brechin, 69.

6. See, e.g., Davies and Young, *Tales from the Dark Side of the City*; and "Grey Goldmine."

7. For more on the street-level changes in San Francisco, see Bloomfield, "History of the California Historical Society's New Mission Street Neighborhood."

8. "Street Homelessness." See also "Counterpoints: An Atlas of Displacement and Resistance."

9. Gee, "San Francisco or Mumbai?"

10. H. W. Turner published a detailed geological survey of the Silver Peak area in July 1909. In beautiful prose, Turner extolled the geological variety within what he described as "slopes of cream and pink tuffs, and little hillocks of a bright brick red." Turner, "Contribution to the Geology of the Silver Peak Quadrangle, Nevada," 228.

11. Lambert, "Breakdown of Raw Materials in Tesla's Batteries and Possible Breaknecks."

12. Bullis, "Lithium-Ion Battery."

13. "Chinese Lithium Giant Agrees to Three-Year Pact to Supply Tesla."

14. Wald, "Tesla Is a Battery Business."

15. Scheyder, "Tesla Expects Global Shortage."

16. Wade, "Tesla's Electric Cars Aren't as Green."

17. Business Council for Sustainable Energy, "2019 Sustainable Energy in America Factbook." U.S. Energy Information Administration, "What Is U.S. Electricity Generation by Energy Source?"

18. Whittaker et al., *AI Now Report 2018*.

19. Parikka, *Geology of Media*, vii–viii; McLuhan, *Understanding Media*.

20. Ely, "Life Expectancy of Electronics."

21. Sandro Mezzadra and Brett Neilson use the term "extractivism" to name the relation between different forms of extractive operations in contemporary capitalism, which we see repeated in the context of the AI industry. Mezzadra and Neilson, "Multiple Frontiers of Extraction."

22. Nassar et al., "Evaluating the Mineral Commodity Supply Risk of the US Manufacturing Sector."

23. Mumford, *Technics and Civilization*, 74.

24. See, e.g., Ayogu and Lewis, "Conflict Minerals."

25. Burke, "Congo Violence Fuels Fears of Return to 90s Bloodbath."

26. "Congo's Bloody Coltan."

27. "Congo's Bloody Coltan."

28. "Transforming Intel's Supply Chain with Real-Time Analytics."

29. See, e.g., an open letter from seventy signatories that criticizes the limitations of the so-called conflict-free certification process: "An Open Letter."

30. "Responsible Minerals Policy and Due Diligence."

31. In *The Elements of Power,* David S. Abraham describes the invisible networks of rare metals traders in global electronics supply chains: "The network to get rare metals from the mine to your laptop travels through a murky network of traders, processors, and component manufacturers. Traders are the middlemen who do more than buy and sell rare metals: they help to regulate information and are the hidden link that helps in navigating the network between metals plants and the components in our laptops" (89).

32. "Responsible Minerals Sourcing."

33. Liu, "Chinese Mining Dump."

34. "Bayan Obo Deposit."

35. Maughan, "Dystopian Lake Filled by the World's Tech Lust."

36. Hird, "Waste, Landfills, and an Environmental Ethics of Vulnerability," 105.

37. Abraham, *Elements of Power,* 175.

38. Abraham, 176.

39. Simpson, "Deadly Tin Inside Your Smartphone."

40. Hodal, "Death Metal."

41. Hodal.

42. Tully, "Victorian Ecological Disaster."

43. Starosielski, *Undersea Network,* 34.

44. See Couldry and Mejías, *Costs of Connection,* 46.

45. Couldry and Mejías, 574.

46. For a superb account of the history of undersea cables, see Starosielski, *Undersea Network.*

47. Dryer, "Designing Certainty," 45.

48. Dryer, 46.

49. Dryer, 266–68.

50. More people are now drawing attention to this problem—including researchers at AI Now. See Dobbe and Whittaker, "AI and Climate Change."

51. See, as an example of early scholarship in this area, Ensmenger, "Computation, Materiality, and the Global Environment."

52. Hu, *Prehistory of the Cloud,* 146.

53. Jones, "How to Stop Data Centres from Gobbling Up the World's Electricity." Some progress has been made toward mitigating these concerns through greater energy efficiency practices, but significant long-term challenges remain. Masanet et al., "Recalibrating Global Data Center Energy-Use Estimates."

54. Belkhir and Elmeligi, "Assessing ICT Global Emissions Footprint"; Andrae and Edler, "On Global Electricity Usage."

55. Strubell, Ganesh, and McCallum, "Energy and Policy Considerations for Deep Learning in NLP."

56. Strubell, Ganesh, and McCallum.

57. Sutton, "Bitter Lesson."

58. "AI and Compute."

59. Cook et al., *Clicking Clean.*

60. Ghaffary, "More Than 1,000 Google Employees Signed a Letter." See also "Apple Commits to Be 100 Percent Carbon Neutral"; Harrabin, "Google Says Its Carbon Footprint Is Now Zero"; Smith, "Microsoft Will Be Carbon Negative by 2030."

61. "Powering the Cloud."

62. "Powering the Cloud."

63. "Powering the Cloud."

64. Hogan, "Data Flows and Water Woes."

65. "Off Now."

66. Carlisle, "Shutting Off NSA's Water Gains Support."

67. Materiality is a complex concept, and there is a lengthy literature that contends with it in such fields as STS, anthropology, and media studies. In one sense, materiality refers to what Leah Lievrouw describes as "the physical character and existence of objects and artifacts that makes them useful and usable for certain purposes under particular conditions." Lievrouw quoted in Gillespie, Boczkowski, and Foot, *Media Technologies,* 25. But as Diana Coole and Samantha Frost write, "Materiality is always something more than 'mere' matter: an excess, force, vitality, relationality, or difference that renders matter active, self-creative, productive, unproductive." Coole and Frost, *New Materialisms,* 9.

68. United Nations Conference on Trade and Development, *Review of Maritime Transport, 2017.*

69. George, *Ninety Percent of Everything,* 4.

70. Schlanger, "If Shipping Were a Country."

71. Vidal, "Health Risks of Shipping Pollution."

72. "Containers Lost at Sea—2017 Update."

73. Adams, "Lost at Sea."

74. Mumford, *Myth of the Machine.*

75. Labban, "Deterritorializing Extraction." For an expansion on this idea, see Arboleda, *Planetary Mine.*

76. Ananny and Crawford, "Seeing without Knowing."

2

Labor

1. Wilson, "Amazon and Target Race."

2. Lingel and Crawford, "Alexa, Tell Me about Your Mother."

3. Federici, *Wages against Housework*; Gregg, *Counterproductive.*

4. In *The Utopia of Rules,* David Graeber details the sense of loss experienced by white-collar workers who now have to enter data into the decision-making systems that have replaced specialist administrative support staff in most professional workplaces.

5. Smith, *Wealth of Nations,* 4–5.

6. Marx and Engels, *Marx-Engels Reader,* 479. Marx expanded on this notion of the worker as an "appendage" in *Capital,* vol. 1: "In handicrafts and manufacture, the worker makes use of a tool; in the factory, the machine makes use of him. There the movements of the instrument of labor proceed from him, here it is the movements of the machine that he must follow. In manufacture the workers are parts of a living mechanism. In the factory we have a lifeless mechanism which is independent of the workers, who are incorporated into it as its living appendages." Marx, *Das Kapital,* 548–49.

7. Luxemburg, "Practical Economies," 444.

8. Thompson, "Time, Work-Discipline, and Industrial Capitalism."

9. Thompson, 88–90.

10. Werrett, "Potemkin and the Panopticon," 6.

11. See, e.g., Cooper, "Portsmouth System of Manufacture."

12. Foucault, *Discipline and Punish;* Horne and Maly, *Inspection House.*

13. Mirzoeff, *Right to Look,* 58.

14. Mirzoeff, 55.

15. Mirzoeff, 56.

16. Gray and Suri, *Ghost Work.*

17. Irani, "Hidden Faces of Automation."

18. Yuan, "How Cheap Labor Drives China's A.I. Ambitions"; Gray and Suri, "Humans Working behind the AI Curtain."

19. Berg et al., *Digital Labour Platforms.*

20. Roberts, *Behind the Screen;* Gillespie, *Custodians of the Internet,* 111–40.

21. Silberman et al., "Responsible Research with Crowds."

22. Silberman et al.

23. Huet, "Humans Hiding behind the Chatbots."

24. Huet.

25. See Sadowski, "Potemkin AI."

26. Taylor, "Automation Charade."

27. Taylor.

28. Gray and Suri, *Ghost Work.*

29. Standage, *Turk,* 23.

30. Standage, 23.

31. See, e.g., Aytes, "Return of the Crowds," 80.

32. Irani, "Difference and Dependence among Digital Workers," 225.

33. Pontin, "Artificial Intelligence."

34. Menabrea and Lovelace, "Sketch of the Analytical Engine."

35. Babbage, *On the Economy of Machinery and Manufactures*, 39–43.

36. Babbage evidently acquired an interest in quality-control processes while trying (vainly) to establish a reliable supply chain for the components of his calculating engines.

37. Schaffer, "Babbage's Calculating Engines and the Factory System," 280.

38. Taylor, *People's Platform*, 42.

39. Katz and Krueger, "Rise and Nature of Alternative Work Arrangements."

40. Rehmann, "Taylorism and Fordism in the Stockyards," 26.

41. Braverman, *Labor and Monopoly Capital*, 56, 67; Specht, *Red Meat Republic*.

42. Taylor, *Principles of Scientific Management*.

43. Marx, *Poverty of Philosophy*, 22.

44. Qiu, Gregg, and Crawford, "Circuits of Labour"; Qiu, *Goodbye iSlave*.

45. Markoff, "Skilled Work, without the Worker."

46. Guendelsberger, *On the Clock*, 22.

47. Greenhouse, "McDonald's Workers File Wage Suits."

48. Greenhouse.

49. Mayhew and Quinlan, "Fordism in the Fast Food Industry."

50. Ajunwa, Crawford, and Schultz, "Limitless Worker Surveillance."

51. Mikel, "WeWork Just Made a Disturbing Acquisition."

52. Mahdawi, "Domino's 'Pizza Checker' Is Just the Beginning."

53. Wajcman, "How Silicon Valley Sets Time."

54. Wajcman, 1277.

55. Gora, Herzog, and Tripathi, "Clock Synchronization."

56. Eglash, "Broken Metaphor," 361.

57. Kemeny and Kurtz, "Dartmouth Timesharing," 223.

58. Eglash, "Broken Metaphor," 364.

59. Brewer, "Spanner, TrueTime."

60. Corbett et al., "Spanner," 14, cited in House, "Synchronizing Uncertainty," 124.

61. Galison, *Einstein's Clocks, Poincaré's Maps*, 104.

62. Galison, 112.

63. Colligan and Linley, "Media, Technology, and Literature," 246.

64. Carey, "Technology and Ideology."

65. Carey, 13.

66. This contrasts with what Foucault called the "microphysics of power" to describe how institutions and apparatuses create particular logics and forms of validity. Foucault, *Discipline and Punish*, 26.

67. Spargo, *Syndicalism, Industrial Unionism, and Socialism*.

68. Personal conversation with the author at an Amazon fulfillment center tour, Robbinsville, N.J., October 8, 2019.

69. Muse, "Organizing Tech."

70. Abdi Muse, personal conversation with the author, October 2, 2019.

71. Gurley, "60 Amazon Workers Walked Out."

72. Muse quoted in *Organizing Tech.*

73. Desai quoted in *Organizing Tech.*

74. Estreicher and Owens, "Labor Board Wrongly Rejects Employee Access to Company Email."

75. This observation comes from conversations with various labor organizers, tech workers, and researchers, including Astra Taylor, Dan Greene, Bo Daley, and Meredith Whittaker.

76. Kerr, "Tech Workers Protest in SF."

3
Data

1. National Institute of Standards and Technology (NIST), "Special Database 32—Multiple Encounter Dataset (MEDS)."

2. Russell, *Open Standards and the Digital Age.*

3. Researchers at NIST (then the National Bureau of Standards, NBS) began working on the first version of the FBI's Automated Fingerprint Identification System in the late 1960s. See Garris and Wilson, "NIST Biometrics Evaluations and Developments," 1.

4. Garris and Wilson, 1.

5. Garris and Wilson, 12.

6. Sekula, "Body and the Archive," 17.

7. Sekula, 18–19.

8. Sekula, 17.

9. See, e.g., Grother et al., "2017 IARPA Face Recognition Prize Challenge (FRPC)."

10. See, e.g., Ever AI, "Ever AI Leads All US Companies."

11. Founds et al., "NIST Special Database 32."

12. Curry et al., "NIST Special Database 32 Multiple Encounter Dataset I (MEDS-I)," 8.

13. See, e.g., Jaton, "We Get the Algorithms of Our Ground Truths."

14. Nilsson, *Quest for Artificial Intelligence,* 398.

15. "ImageNet Large Scale Visual Recognition Competition (ILSVRC)."

16. In the late 1970s, Ryszard Michalski wrote an algorithm based on symbolic variables and logical rules. This language was popular in the 1980s and 1990s, but as the rules of decision-making and qualification became more complex, the language became less usable. At the same moment, the potential of using large training sets triggered a shift from this conceptual clus-

tering to contemporary machine learning approaches. Michalski, "Pattern Recognition as Rule-Guided Inductive Inference."

17. Bush, "As We May Think."

18. Light, "When Computers Were Women"; Hicks, *Programmed Inequality.*

19. As described in Russell and Norvig, *Artificial Intelligence,* 546.

20. Li, "Divination Engines," 143.

21. Li, 144.

22. Brown and Mercer, "Oh, Yes, Everything's Right on Schedule, Fred."

23. Lem, "First Sally (A), or Trurl's Electronic Bard," 199.

24. Lem, 199.

25. Brown and Mercer, "Oh, Yes, Everything's Right on Schedule, Fred."

26. Marcus, Marcinkiewicz, and Santorini, "Building a Large Annotated Corpus of English."

27. Klimt and Yang, "Enron Corpus."

28. Wood, Massey, and Brownell, "FERC Order Directing Release of Information," 12.

29. Heller, "What the Enron Emails Say about Us."

30. Baker et al., "Research Developments and Directions in Speech Recognition."

31. I have participated in early work to address this gap. See, e.g., Gebru et al., "Datasheets for Datasets." Other researchers have also sought to address this problem for AI models; see Mitchell et al., "Model Cards for Model Reporting"; Raji and Buolamwini, "Actionable Auditing."

32. Phillips, Rauss, and Der, "FERET (Face Recognition Technology) Recognition Algorithm Development and Test Results," 9.

33. Phillips, Rauss, and Der, 61.

34. Phillips, Rauss, and Der, 12.

35. See Aslam, "Facebook by the Numbers (2019)"; and "Advertising on Twitter."

36. Fei-Fei Li, as quoted in Gershgorn, "Data That Transformed AI Research."

37. Deng et al., "ImageNet."

38. Gershgorn, "Data That Transformed AI Research."

39. Gershgorn.

40. Markoff, "Seeking a Better Way to Find Web Images."

41. Hernandez, "CU Colorado Springs Students Secretly Photographed."

42. Zhang et al., "Multi-Target, Multi-Camera Tracking by Hierarchical Clustering."

43. Sheridan, "Duke Study Recorded Thousands of Students' Faces."

44. Harvey and LaPlace, "Brainwash Dataset."

45. Locker, "Microsoft, Duke, and Stanford Quietly Delete Databases."

46. Murgia and Harlow, "Who's Using Your Face?" When the *Financial*

Times exposed the contents of this dataset, Microsoft removed the set from the internet, and a spokesperson for Microsoft claimed simply that it was removed "because the research challenge is over." Locker, "Microsoft, Duke, and Stanford Quietly Delete Databases."

47. Franceschi-Bicchierai, "Redditor Cracks Anonymous Data Trove."

48. Tockar, "Riding with the Stars."

49. Crawford and Schultz, "Big Data and Due Process."

50. Franceschi-Bicchierai, "Redditor Cracks Anonymous Data Trove."

51. Nilsson, *Quest for Artificial Intelligence*, 495.

52. And, as Geoff Bowker famously reminds us, "Raw data is both an oxymoron and a bad idea; to the contrary, data should be cooked with care." Bowker, *Memory Practices in the Sciences*, 184–85.

53. Fourcade and Healy, "Seeing Like a Market," 13, emphasis added.

54. Meyer and Jepperson, "'Actors' of Modern Society."

55. Gitelman, *"Raw Data" Is an Oxymoron*, 3.

56. Many scholars have looked closely at the work these metaphors do. Media studies professors Cornelius Puschmann and Jean Burgess analyzed the common data metaphors and noted two widespread categories: data "as a natural force to be controlled and [data] as a resource to be consumed." Puschmann and Burgess, "Big Data, Big Questions," abstract. Researchers Tim Hwang and Karen Levy suggest that describing data as "the new oil" carries connotations of being costly to acquire but also suggests the possibility of "big payoffs for those with the means to extract it." Hwang and Levy, "'The Cloud' and Other Dangerous Metaphors."

57. Stark and Hoffmann, "Data Is the New What?"

58. Media scholars Nick Couldry and Ulises Mejías call this "data colonialism," which is steeped in the historical, predatory practices of colonialism but married to (and obscured by) contemporary computing methods. However, as other scholars have shown, this terminology is double-edged because it can occlude the real and ongoing harms of colonialism. Couldry and Mejías, "Data Colonialism"; Couldry and Mejías, *Costs of Connection*; Segura and Waisbord, "Between Data Capitalism and Data Citizenship."

59. They refer to this form of capital as "ubercapital." Fourcade and Healy, "Seeing Like a Market," 19.

60. Sadowski, "When Data Is Capital," 8.

61. Sadowski, 9.

62. Here I'm drawing from a history of human subjects review and large-scale data studies coauthored with Jake Metcalf. See Metcalf and Crawford, "Where Are Human Subjects in Big Data Research?"

63. "Federal Policy for the Protection of Human Subjects."

64. See Metcalf and Crawford, "Where Are Human Subjects in Big Data Research?"

65. Seo et al., "Partially Generative Neural Networks." Jeffrey Branting-ham, one of the authors, is also a co-founder of the controversial predictive policing company PredPol. See Winston and Burrington, "A Pioneer in Pre-dictive Policing."

66. "CalGang Criminal Intelligence System."

67. Libby, "Scathing Audit Bolsters Critics' Fears."

68. Hutson, "Artificial Intelligence Could Identify Gang Crimes."

69. Hoffmann, "Data Violence and How Bad Engineering Choices Can Damage Society."

70. Weizenbaum, *Computer Power and Human Reason,* 266.

71. Weizenbaum, 275–76.

72. Weizenbaum, 276.

73. For more on the history of extraction of data and insights from mar-ginalized communities, see Costanza-Chock, *Design Justice;* and D'Ignazio and Klein, *Data Feminism.*

74. Revell, "Google DeepMind's NHS Data Deal 'Failed to Comply.'"

75. "Royal Free–Google DeepMind Trial Failed to Comply."

4
Classification

1. Fabian, *Skull Collectors.*

2. Gould, *Mismeasure of Man,* 83.

3. Kolbert, "There's No Scientific Basis for Race."

4. Keel, "Religion, Polygenism and the Early Science of Human Origins."

5. Thomas, *Skull Wars.*

6. Thomas, 85.

7. Kendi, "History of Race and Racism in America."

8. Gould, *Mismeasure of Man,* 88.

9. Mitchell, "Fault in His Seeds."

10. Horowitz, "Why Brain Size Doesn't Correlate with Intelligence."

11. Mitchell, "Fault in His Seeds."

12. Gould, *Mismeasure of Man,* 58.

13. West, "Genealogy of Modern Racism," 91.

14. Bouche and Rivard, "America's Hidden History."

15. Bowker and Star, *Sorting Things Out,* 319.

16. Bowker and Star, 319.

17. Nedlund, "Apple Card Is Accused of Gender Bias"; Angwin et al., "Ma-chine Bias"; Angwin et al., "Dozens of Companies Are Using Facebook to Exclude."

18. Dougherty, "Google Photos Mistakenly Labels Black People 'Goril-las'"; Perez, "Microsoft Silences Its New A.I. Bot Tay"; McMillan, "It's Not

You, It's It"; Sloane, "Online Ads for High-Paying Jobs Are Targeting Men More Than Women."

19. See Benjamin, *Race after Technology*; and Noble, *Algorithms of Oppression*.

20. Greene, "Science May Have Cured Biased AI"; Natarajan, "Amazon and NSF Collaborate to Accelerate Fairness in AI Research."

21. Dastin, "Amazon Scraps Secret AI Recruiting Tool."

22. Dastin.

23. This is part of a larger trend toward automating aspects of hiring. For a detailed account, see Ajunwa and Greene, "Platforms at Work."

24. There are several superb accounts of the history of inequality and discrimination in computation. These are a few that have informed my thinking on these issues: Hicks, *Programmed Inequality*; McIlwain, *Black Software*; Light, "When Computers Were Women"; and Ensmenger, *Computer Boys Take Over.*

25. Cetina, *Epistemic Cultures*, 3.

26. Merler et al., "Diversity in Faces."

27. Buolamwini and Gebru, "Gender Shades"; Raj et al. "Saving Face."

28. Merler et al., "Diversity in Faces."

29. "YFCC100M Core Dataset."

30. Merler et al., "Diversity in Faces," 1.

31. There are many excellent books on these issues, but in particular, see Roberts, *Fatal Invention*, 18–41; and Nelson, *Social Life of DNA*, 43. See also Tishkoff and Kidd, "Implications of Biogeography."

32. Browne, "Digital Epidermalization," 135.

33. Benthall and Haynes, "Racial Categories in Machine Learning."

34. Mitchell, "Need for Biases in Learning Generalizations."

35. Dietterich and Kong, "Machine Learning Bias, Statistical Bias."

36. Domingos, "Useful Things to Know about Machine Learning."

37. *Maddox v. State*, 32 Ga. 5S7, 79 Am. Dec. 307; *Pierson v. State*, 18 Tex. App. 55S; *Hinkle v. State*, 94 Ga. 595, 21 S. E. 601.

38. Tversky and Kahneman, "Judgment under Uncertainty."

39. Greenwald and Krieger, "Implicit Bias," 951.

40. Fellbaum, *WordNet*, xviii. Below I am drawing on research into ImageNet conducted with Trevor Paglen. See Crawford and Paglen, "Excavating AI."

41. Fellbaum, xix.

42. Nelson and Kucera, *Brown Corpus Manual.*

43. Borges, "The Analytical Language of John Wilkins."

44. These are some of the categories that have now been deleted entirely from ImageNet as of October 1, 2020.

45. See Keyes, "Misgendering Machines."

46. Drescher, "Out of DSM."

47. See Bayer, *Homosexuality and American Psychiatry.*

48. Keyes, "Misgendering Machines."

49. Hacking, "Making Up People," 23.

50. Bowker and Star, *Sorting Things Out,* 196.

51. This is drawn from Lakoff, *Women, Fire, and Dangerous Things.*

52. ImageNet Roulette was one of the outputs of a multiyear research collaboration between the artist Trevor Paglen and me, in which we studied the underlying logic of multiple benchmark training sets in AI. ImageNet Roulette, led by Paglen and produced by Leif Ryge, was an app that allowed people to interact with a neural net trained on the "person" category of ImageNet. People could upload images of themselves — or news images or historical photographs — to see how ImageNet would label them. People could also see how many of the labels are bizarre, racist, misogynist, and otherwise problematic. The app was designed to show people these concerning labels while warning them in advance of the potential results. All uploaded image data were immediately deleted on processing. See Crawford and Paglen, "Excavating AI."

53. Yang et al., "Towards Fairer Datasets," paragraph 4.2.

54. Yang et al., paragraph 4.3.

55. Markoff, "Seeking a Better Way to Find Web Images."

56. Browne, *Dark Matters,* 114.

57. Scheuerman et al., "How We've Taught Algorithms to See Identity."

58. UTKFace Large Scale Face Dataset, https://susanqq.github.io/UTK Face.

59. Bowker and Star, *Sorting Things Out,* 197.

60. Bowker and Star, 198.

61. Edwards and Hecht, "History and the Technopolitics of Identity," 627.

62. Haraway, *Modest_Witness@Second_Millennium,* 234.

63. Stark, "Facial Recognition Is the Plutonium of AI," 53.

64. In order of the examples, see Wang and Kosinski, "Deep Neural Networks Are More Accurate than Humans"; Wu and Zhang, "Automated Inference on Criminality Using Face Images"; and Angwin et al., "Machine Bias."

65. Agüera y Arcas, Mitchell, and Todorov, "Physiognomy's New Clothes."

66. Nielsen, *Disability History of the United States;* Kafer, *Feminist, Queer, Crip;* Siebers, *Disability Theory.*

67. Whittaker et al., "Disability, Bias, and AI."

68. Hacking, "Kinds of People," 289.

69. Bowker and Star, *Sorting Things Out,* 31.

70. Bowker and Star, 6.

71. Eco, *Infinity of Lists.*

72. Douglass, "West India Emancipation."

5
Affect

1. Particular thanks to Alex Campolo, who was my research assistant and interlocutor for this chapter, and for his research into Ekman and the history of emotions.

2. "Emotion Detection and Recognition"; Schwartz, "Don't Look Now."

3. Ohtake, "Psychologist Paul Ekman Delights at Exploratorium."

4. Ekman, *Emotions Revealed*, 7.

5. For an overview of researchers who have found flaws in the claim that emotional expressions are universal and can be predicted by AI, see Heaven, "Why Faces Don't Always Tell the Truth."

6. Barrett et al., "Emotional Expressions Reconsidered."

7. Nilsson, "How AI Helps Recruiters."

8. Sánchez-Monedero and Dencik, "Datafication of the Workplace," 48; Harwell, "Face-Scanning Algorithm."

9. Byford, "Apple Buys Emotient."

10. Molnar, Robbins, and Pierson, "Cutting Edge."

11. Picard, "Affective Computing Group."

12. "Affectiva Human Perception AI Analyzes Complex Human States."

13. Schwartz, "Don't Look Now."

14. See, e.g., Nilsson, "How AI Helps Recruiters."

15. "Face: An AI Service That Analyzes Faces in Images."

16. "Amazon Rekognition Improves Face Analysis"; "Amazon Rekognition — Video and Image."

17. Barrett et al., "Emotional Expressions Reconsidered," 1.

18. Sedgwick, Frank, and Alexander, *Shame and Its Sisters*, 258.

19. Tomkins, *Affect Imagery Consciousness.*

20. Tomkins.

21. Leys, *Ascent of Affect*, 18.

22. Tomkins, *Affect Imagery Consciousness*, 23.

23. Tomkins, 23.

24. Tomkins, 23.

25. For Ruth Leys, this "radical dissociation between feeling and cognition" is the major reason for its attractiveness to theorists in the humanities, most notably Eve Kosofsky Sedgwick, who wants to revalorize our experiences of error or confusion into new forms of freedom. Leys, *Ascent of Affect*, 35; Sedgwick, *Touching Feeling.*

26. Tomkins, *Affect Imagery Consciousness*, 204.

27. Tomkins, 206; Darwin, *Expression of the Emotions*; Duchenne (de Boulogne), *Mécanisme de la physionomie humaine.*

28. Tomkins, 243, quoted in Leys, *Ascent of Affect*, 32.

29. Tomkins, *Affect Imagery Consciousness*, 216.

30. Ekman, *Nonverbal Messages*, 45.

31. Tuschling, "Age of Affective Computing," 186.

32. Ekman, *Nonverbal Messages*, 45.

33. Ekman, 46.

34. Ekman, 46.

35. Ekman, 46.

36. Ekman, 46.

37. Ekman, 46.

38. Ekman and Rosenberg, *What the Face Reveals*, 375.

39. Tomkins and McCarter, "What and Where Are the Primary Affects?"

40. Russell, "Is There Universal Recognition of Emotion from Facial Expression?" 116.

41. Leys, *Ascent of Affect*, 93.

42. Ekman and Rosenberg, *What the Face Reveals*, 377.

43. Ekman, Sorenson, and Friesen, "Pan-Cultural Elements in Facial Displays of Emotion," 86, 87.

44. Ekman and Friesen, "Constants across Cultures in the Face and Emotion," 128.

45. Aristotle, *Categories*, 70b8–13, 527.

46. Aristotle, 805a, 27–30, 87.

47. It would be difficult to overstate the influence of this work, which has since fallen into disrepute: by 1810 it went through sixteen German and twenty English editions. Graham, "Lavater's Physiognomy in England," 561.

48. Gray, *About Face*, 342.

49. Courtine and Haroche, *Histoire du visage*, 132.

50. Ekman, "Duchenne and Facial Expression of Emotion."

51. Duchenne (de Boulogne), *Mécanisme de la physionomie humaine.*

52. Clarac, Massion, and Smith, "Duchenne, Charcot and Babinski," 362–63.

53. Delaporte, *Anatomy of the Passions*, 33.

54. Delaporte, 48–51.

55. Daston and Galison, *Objectivity.*

56. Darwin, *Expression of the Emotions in Man and Animals*, 12, 307.

57. Leys, *Ascent of Affect*, 85; Russell, "Universal Recognition of Emotion," 114.

58. Ekman and Friesen, "Nonverbal Leakage and Clues to Deception," 93.

59. Pontin, "Lie Detection."

60. Ekman and Friesen, "Nonverbal Leakage and Clues to Deception," 94. In a footnote, Ekman and Friesen explained: "Our own research and the evidence from the neurophysiology of visual perception strongly suggest

that micro-expressions that are as short as one motion-picture frame (1/50 of a second) can be perceived. That these micro-expressions are not usually seen must depend upon their being embedded in other expressions which distract attention, their infrequency, or some learned perceptual habit of ignoring fast facial expressions."

61. Ekman, Sorenson, and Friesen, "Pan-Cultural Elements in Facial Displays of Emotion," 87.

62. Ekman, Friesen, and Tomkins, "Facial Affect Scoring Technique," 40.

63. Ekman, *Nonverbal Messages,* 97.

64. Ekman, 102.

65. Ekman and Rosenberg, *What the Face Reveals.*

66. Ekman, *Nonverbal Messages,* 105.

67. Ekman, 169.

68. Eckman, 106; Aleksander, *Artificial Vision for Robots.*

69. "Magic from Invention."

70. Bledsoe, "Model Method in Facial Recognition."

71. Molnar, Robbins, and Pierson, "Cutting Edge."

72. Kanade, *Computer Recognition of Human Faces.*

73. Kanade, 16.

74. Kanade, Cohn, and Tian, "Comprehensive Database for Facial Expression Analysis," 6.

75. See Kanade, Cohn, and Tian; Lyons et al., "Coding Facial Expressions with Gabor Wavelets"; and Goeleven et al., "Karolinska Directed Emotional Faces."

76. Lucey et al., "Extended Cohn-Kanade Dataset (CK+)."

77. McDuff et al., "Affectiva-MIT Facial Expression Dataset (AM-FED)."

78. McDuff et al.

79. Ekman and Friesen, *Facial Action Coding System (FACS).*

80. Foreman, "Conversation with: Paul Ekman"; Taylor, "2009 Time 100"; Paul Ekman Group.

81. Weinberger, "Airport Security," 413.

82. Halsey, "House Member Questions $900 Million TSA 'SPOT' Screening Program."

83. Ekman, "Life's Pursuit"; Ekman, *Nonverbal Messages,* 79–81.

84. Mead, "Review of *Darwin and Facial Expression,*" 209.

85. Tomkins, *Affect Imagery Consciousness,* 216.

86. Mead, "Review of *Darwin and Facial Expression,*" 212. See also Fridlund, "Behavioral Ecology View of Facial Displays." Ekman later conceded to many of Mead's points. See Ekman, "Argument for Basic Emotions"; Ekman, *Emotions Revealed;* and Ekman, "What Scientists Who Study Emotion Agree About." Ekman also had his defenders. See Cowen et al., "Mapping the Pas-

sions"; and Elfenbein and Ambady, "Universality and Cultural Specificity of Emotion Recognition."

87. Fernández-Dols and Russell, *Science of Facial Expression,* 4.

88. Gendron and Barrett, *Facing the Past,* 30.

89. Vincent, "AI 'Emotion Recognition' Can't Be Trusted.'" Disability studies scholars have also noted that assumptions about how biology and bodies function can also raise concerns around bias, especially when automated through technology. See Whittaker et al., "Disability, Bias, and AI."

90. Izard, "Many Meanings/Aspects of Emotion."

91. Leys, *Ascent of Affect,* 22.

92. Leys, 92.

93. Leys, 94.

94. Leys, 94.

95. Barrett, "Are Emotions Natural Kinds?" 28.

96. Barrett, 30.

97. See, e.g., Barrett et al., "Emotional Expressions Reconsidered."

98. Barrett et al., 40.

99. Kappas, "Smile When You Read This," 39, emphasis added.

100. Kappas, 40.

101. Barrett et al., 46.

102. Barrett et al., 47–48.

103. Barrett et al., 47, emphasis added.

104. Apelbaum, "One Thousand and One Nights."

105. See, e.g., Hoft, "Facial, Speech and Virtual Polygraph Analysis."

106. Rhue, "Racial Influence on Automated Perceptions of Emotions."

107. Barrett et al., "Emotional Expressions Reconsidered,"48.

108. See, e.g., Connor, "Chinese School Uses Facial Recognition"; and Du and Maki, "AI Cameras That Can Spot Shoplifters."

6
State

1. NOFORN stands for Not Releasable to Foreign Nationals. "Use of the 'Not Releasable to Foreign Nationals' (NOFORN) Caveat."

2. The Five Eyes is a global intelligence alliance comprising Australia, Canada, New Zealand, the United Kingdom, and the United States. "Five Eyes Intelligence Oversight and Review Council."

3. Galison, "Removing Knowledge," 229.

4. Risen and Poitras, "N.S.A. Report Outlined Goals for More Power"; Müller-Maguhn et al., "The NSA Breach of Telekom and Other German Firms."

5. FOXACID is software developed by the Office of Tailored Access Operations, now Computer Network Operations, a cyberwarfare intelligence-gathering unit of the NSA.

6. Schneier, "Attacking Tor." Document available at "NSA Phishing Tactics and Man in the Middle Attacks."

7. Swinhoe, "What Is Spear Phishing?"

8. "Strategy for Surveillance Powers."

9. Edwards, *Closed World.*

10. Edwards.

11. Edwards, 198.

12. Mbembé, *Necropolitics,* 82.

13. Bratton, *Stack,* 151.

14. For an excellent account of the history of the internet in the United States, see Abbate, *Inventing the Internet.*

15. SHARE Foundation, "Serbian Government Is Implementing Unlawful Video Surveillance."

16. Department of International Cooperation Ministry of Science and Technology, "Next Generation Artificial Intelligence Development Plan."

17. Chun, *Control and Freedom;* Hu, *Prehistory of the Cloud,* 87–88.

18. Cave and ÓhÉigeartaigh, "AI Race for Strategic Advantage."

19. Markoff, "Pentagon Turns to Silicon Valley for Edge."

20. Brown, *Department of Defense Annual Report.*

21. Martinage, "Toward a New Offset Strategy," 5–16.

22. Carter, "Remarks on 'the Path to an Innovative Future for Defense'"; Pellerin, "Deputy Secretary."

23. The origins of U.S. military offsets can be traced back to December 1952, when the Soviet Union had almost ten times more conventional military divisions than the United States. President Dwight Eisenhower turned to nuclear deterrence as a way to "offset" these odds. The strategy involved not only the threat of the retaliatory power of the U.S. nuclear forces but also accelerating the growth of the U.S. weapons stockpile, as well as developing long-range jet bombers, the hydrogen bomb, and eventually intercontinental ballistic missiles. It also included increased reliance on espionage, sabotage, and covert operations. In the 1970s and 1980s, U.S. military strategy turned to computational advances in analytics and logistics, building on the influence of such military architects as Robert McNamara in search of military supremacy. This Second Offset could be seen in military engagements like Operation Desert Storm during the Gulf War in 1991, where reconnaissance, suppression of enemy defenses, and precision-guided munitions dominated how the United States not only fought the war but thought and spoke about it. Yet as Russia and China began to adopt these capacities and deploy digi-

tal networks for warfare, anxiety grew to reestablish a new kind of strategic advantage. See McNamara and Blight, *Wilson's Ghost.*

24. Pellerin, "Deputy Secretary."

25. Gellman and Poitras, "U.S., British Intelligence Mining Data."

26. Deputy Secretary of Defense to Secretaries of the Military Departments et al.

27. Deputy Secretary of Defense to Secretaries of the Military Departments et al.

28. Michel, *Eyes in the Sky,* 134.

29. Michel, 135.

30. Cameron and Conger, "Google Is Helping the Pentagon Build AI for Drones."

31. For example, Gebru et al., "Fine-Grained Car Detection for Visual Census Estimation."

32. Fang, "Leaked Emails Show Google Expected Lucrative Military Drone AI Work."

33. Bergen, "Pentagon Drone Program Is Using Google AI."

34. Shane and Wakabayashi, "'Business of War.'"

35. Smith, "Technology and the US Military."

36. When the JEDI contract was ultimately awarded to Microsoft, Brad Smith, the president of Microsoft, explained that the reason that Microsoft won the contract was that it was seen "not just as a sales opportunity, but really, a very large-scale engineering project." Stewart and Carlson, "President of Microsoft Says It Took Its Bid."

37. Pichai, "AI at Google."

38. Pichai. Project Maven was subsequently picked up by Anduril Industries, a secretive tech startup founded by Oculus Rift's Palmer Luckey. Fang, "Defense Tech Startup."

39. Whittaker et al., *AI Now Report 2018.*

40. Schmidt quoted in Scharre et al., "Eric Schmidt Keynote Address."

41. As Suchman notes, "'Killing people correctly' under the laws of war requires adherence to the Principle of Distinction and the identification of an imminent threat." Suchman, "Algorithmic Warfare and the Reinvention of Accuracy," n. 18.

42. Suchman.

43. Suchman.

44. Hagendorff, "Ethics of AI Ethics."

45. Brustein and Bergen, "Google Wants to Do Business with the Military."

46. For more on why municipalities should more carefully assess the risks of algorithmic platforms, see Green, *Smart Enough City.*

47. Thiel, "Good for Google, Bad for America."

48. Steinberger, "Does Palantir See Too Much?"

49. Weigel, "Palantir goes to the Frankfurt School."

50. Dilanian, "US Special Operations Forces Are Clamoring to Use Software."

51. "War against Immigrants."

52. Alden, "Inside Palantir, Silicon Valley's Most Secretive Company."

53. Alden, "Inside Palantir, Silicon Valley's Most Secretive Company."

54. Waldman, Chapman, and Robertson, "Palantir Knows Everything about You."

55. Joseph, "Data Company Directly Powers Immigration Raids in Workplace"; Anzilotti, "Emails Show That ICE Uses Palantir Technology to Detain Undocumented Immigrants."

56. Andrew Ferguson, conversation with author, June 21, 2019.

57. Brayne, "Big Data Surveillance." Brayne also notes that the migration of law enforcement to intelligence was occurring even before the shift to predictive analytics, given such court decisions as Terry v. Ohio and Whren v. United States that made it easier for law enforcement to circumvent probable cause and produced a proliferation of pretext stops.

58. Richardson, Schultz, and Crawford, "Dirty Data, Bad Predictions."

59. Brayne, "Big Data Surveillance," 997.

60. Brayne, 997.

61. See, e.g., French and Browne, "Surveillance as Social Regulation."

62. Crawford and Schultz, "AI Systems as State Actors."

63. Cohen, *Between Truth and Power;* Calo and Citron, "Automated Administrative State."

64. "Vigilant Solutions"; Maass and Lipton, "What We Learned."

65. Newman, "Internal Docs Show How ICE Gets Surveillance Help."

66. England, "UK Police's Facial Recognition System."

67. Scott, *Seeing Like a State.*

68. Haskins, "How Ring Transmits Fear to American Suburbs."

69. Haskins, "Amazon's Home Security Company."

70. Haskins.

71. Haskins. "Amazon Requires Police to Shill Surveillance Cameras."

72. Haskins, "Amazon Is Coaching Cops."

73. Haskins.

74. Haskins.

75. Hu, *Prehistory of the Cloud,* 115.

76. Hu, 115.

77. Benson, "'Kill 'Em and Sort It Out Later,'" 17.

78. Hajjar, "Lawfare and Armed Conflicts," 70.

79. Scahill and Greenwald, "NSA's Secret Role in the U.S. Assassination Program."

80. Cole, "'We Kill People Based on Metadata.'"

81. Priest, "NSA Growth Fueled by Need to Target Terrorists."

82. Gibson quoted in Ackerman, "41 Men Targeted but 1,147 People Killed."

83. Tucker, "Refugee or Terrorist?"

84. Tucker.

85. O'Neil, *Weapons of Math Destruction,* 288–326.

86. Fourcade and Healy, "Seeing Like a Market."

87. Eubanks, *Automating Inequality.*

88. Richardson, Schultz, and Southerland, "Litigating Algorithms," 19.

89. Richardson, Schultz, and Southerland, 23.

90. Agre, *Computation and Human Experience,* 240.

91. Bratton, *Stack,* 140.

92. Hu, *Prehistory of the Cloud,* 89.

93. Nakashima and Warrick, "For NSA Chief, Terrorist Threat Drives Passion."

94. Document available at Maass, "Summit Fever."

95. The future of the Snowden archive itself is uncertain. In March 2019, it was announced that the *Intercept*—the publication that Glenn Greenwald established with Laura Poitras and Jeremy Scahill after they shared the Pulitzer Prize for their reporting on the Snowden materials—was no longer going to fund the Snowden archive. Tani, "Intercept Shuts Down Access to Snowden Trove."

Conclusion

1. Silver et al., "Mastering the Game of Go without Human Knowledge."

2. Silver et al., 357.

3. Full talk at the Artificial Intelligence Channel: *Demis Hassabis, DeepMind—Learning from First Principles.* See also Knight, "Alpha Zero's 'Alien' Chess Shows the Power."

4. *Demis Hassabis, DeepMind—Learning from First Principles.*

5. For more on the myths of "magic" in AI, see Elish and boyd, "Situating Methods in the Magic of Big Data and AI."

6. Meredith Broussard notes that playing games has been dangerously conflated with intelligence. She cites the programmer George V. Neville-Neil, who argues: "We have had nearly 50 years of human/computer competition in the game of chess, but does this mean that any of those computers are intelligent? No, it does not—for two reasons. The first is that chess is not a test of intelligence; it is the test of a particular skill—the skill of playing chess. If I could beat a Grandmaster at chess and yet not be able to hand you

the salt at the table when asked, would I be intelligent? The second reason is that thinking chess was a test of intelligence was based on a false cultural premise that brilliant chess players were brilliant minds, more gifted than those around them. Yes, many intelligent people excel at chess, but chess, or any other single skill, does not denote intelligence." Broussard, *Artificial Unintelligence*, 206.

7. Galison, "Ontology of the Enemy."

8. Campolo and Crawford, "Enchanted Determinism."

9. Bailey, "Dimensions of Rhetoric in Conditions of Uncertainty," 30.

10. Bostrom, *Superintelligence*.

11. Bostrom.

12. Strand, "Keyword: Evil," 64–65.

13. Strand, 65.

14. Hardt and Negri, *Assembly*, 116, emphasis added.

15. Wakabayashi, "Google's Shadow Work Force."

16. Quoted in McNeil, "Two Eyes See More Than Nine," 23.

17. On the idea of data as capital, see Sadowski, "When Data Is Capital."

18. Harun Farocki discussed in Paglen, "Operational Images."

19. For a summary, see Heaven, "Why Faces Don't Always Tell the Truth."

20. Nietzsche, *Sämtliche Werke*, 11:506.

21. Wang and Kosinski, "Deep Neural Networks Are More Accurate Than Humans"; Kleinberg et al., "Human Decisions and Machine Predictions"; Crosman, "Is AI a Threat to Fair Lending?"; Seo et al., "Partially Generative Neural Networks."

22. Pugliese, "Death by Metadata."

23. Suchman, "Algorithmic Warfare and the Reinvention of Accuracy."

24. Simmons, "Rekor Software Adds License Plate Reader Technology."

25. Lorde, *Master's Tools*.

26. Schaake, "What Principles Not to Disrupt."

27. Jobin, Ienca, and Vayena, "Global Landscape of AI Ethics Guidelines."

28. Mattern, "Calculative Composition," 572.

29. For more on why AI ethics frameworks are limited in effectiveness, see Crawford et al., *AI Now 2019 Report*.

30. Mittelstadt, "Principles Alone Cannot Guarantee Ethical AI." See also Metcalf, Moss, and boyd, "Owning Ethics."

31. For recent scholarship that addresses important practical steps to do this without replicating forms of extraction and harm, see Costanza-Chock, *Design Justice*.

32. Winner, *The Whale and the Reactor*, 9.

33. Mbembé, *Critique of Black Reason*, 3.

34. Bangstad et al., "Thoughts on the Planetary."

35. Haraway, *Simians, Cyborgs, and Women,* 161.

36. Mohamed, Png, and Isaac, "Decolonial AI," 405.

37. "Race after Technology, Ruha Benjamin."

38. Bangstad et al., "Thoughts on the Planetary."

Coda

1. *Blue Origin's Mission.*

2. *Blue Origin's Mission.*

3. Powell, "Jeff Bezos Foresees a Trillion People."

4. Bezos, *Going to Space to Benefit Earth.*

5. Bezos.

6. Foer, "Jeff Bezos's Master Plan."

7. Foer.

8. "Why Asteroids."

9. Welch, "Elon Musk."

10. Cuthbertson, "Elon Musk Really Wants to 'Nuke Mars.'"

11. Rein, Tamayo, and Vokrouhlicky, "Random Walk of Cars."

12. Gates, "Bezos' Blue Origin Seeks Tax Incentives."

13. Marx, "Instead of Throwing Money at the Moon"; O'Neill, *High Frontier.*

14. "Our Mission."

15. Davis, "Gerard K. O'Neill on Space Colonies."

16. Meadows et al., *Limits to Growth.*

17. Scharmen, *Space Settlements,* 216. In recent years, scholars have suggested that the Club of Rome's models were overly optimistic, underestimating the rapid rate of extraction and resource consumption worldwide and the climate implications of greenhouse gases and industrial waste heat. See Turner, "Is Global Collapse Imminent?"

18. The case for a no-growth model that involves staying on the planet has been made by many academics in the limits to growth movement. See, e.g., Trainer, *Renewable Energy Cannot Sustain a Consumer Society.*

19. Scharmen, *Space Settlements,* 91.

20. One wonders how the Bezos mission would differ had he been inspired instead by the science fiction author Philip K. Dick, who wrote the short story "Autofac" in 1955. In it, human survivors of an apocalyptic war are left on Earth with the "autofacs" — autonomous, self-replicating factory machines. The autofacs had been tasked with producing consumer goods in prewar society but could no longer stop, consuming the planet's resources and threatening the survival of the last people left. The only way to survive was to trick the artificial intelligence machines to fight against each other

over a critical element they need for manufacturing: the rare earth element tungsten. It seems to succeed, and wild vines begin to grow throughout the factories, and farmers can return to the land. Only later do they realize that the autofacs had sought more resources deep in Earth's core and would soon launch thousands of self-replicating "seeds" to mine the rest of the galaxy. Dick, "Autofac."

21. NASA, "Outer Space Treaty of 1967."

22. U.S. Commercial Space Launch Competitiveness Act."

23. Wilson, "Top Lobbying Victories of 2015."

24. Shaer, "Asteroid Miner's Guide to the Galaxy."

25. As Mark Andrejevic writes, "The promise of technological immortality is inseparable from automation, which offers to supplant human limitations at every turn." Andrejevic, *Automated Media*, 1.

26. Reichhardt, "First Photo from Space."

27. See, e.g., Pulitzer Prize–winning journalist Wayne Biddle's account of von Braun as a war criminal who participated in the brutal treatment of slave laborers under the Nazi regime. Biddle, *Dark Side of the Moon.*

28. Grigorieff, "Mittelwerk/Mittelbau/Camp Dora."

29. Ward, *Dr. Space.*

30. Keates, "Many Places Amazon CEO Jeff Bezos Calls Home."

31. Center for Land Use Interpretation, "Figure 2 Ranch, Texas."

Bibliography

Abbate, Janet. *Inventing the Internet*. Cambridge, Mass.: MIT Press, 1999.

Abraham, David S. *The Elements of Power: Gadgets, Guns, and the Struggle for a Sustainable Future in the Rare Metal Age*. New Haven: Yale University Press, 2017.

Achtenberg, Emily. "Bolivia Bets on State-Run Lithium Industry." NACLA, November 15, 2010. https://nacla.org/news/bolivia-bets-state-run-lithium -industry.

Ackerman, Spencer. "41 Men Targeted but 1,147 People Killed: US Drone Strikes — the Facts on the Ground." *Guardian,* November 24, 2014. https:// www.theguardian.com/us-news/2014/nov/24/-sp-us-drone-strikes-kill -1147.

Adams, Guy. "Lost at Sea: On the Trail of Moby-Duck." *Independent,* February 27, 2011. https://www.independent.co.uk/environment/nature/lost -at-sea-on-the-trail-of-moby-duck-2226788.html.

"Advertising on Twitter." Twitter for Business. https://business.twitter.com /en/Twitter-ads-signup.html.

"Affectiva Human Perception AI Analyzes Complex Human States." Affectiva. https://www.affectiva.com/.

Agre, Philip E. *Computation and Human Experience*. Cambridge: Cambridge University Press, 1997.

Agüera y Arcas, Blaise, Margaret Mitchell, and Alexander Todorov. "Physiognomy's New Clothes." *Medium: Artificial Intelligence* (blog), May 7, 2017. https://medium.com/@blaisea/physiognomys-new-clothes-f2d4b59 fdd6a.

"AI and Compute." Open AI, May 16, 2018. https://openai.com/blog/ai-and -compute/.

Alden, William. "Inside Palantir, Silicon Valley's Most Secretive Company."

Buzzfeed News, May 6, 2016, https://www.buzzfeednews.com/article
/williamalden/inside-palantir-silicon-valleys-most-secretive-company.

Ajunwa, Ifeoma, Kate Crawford, and Jason Schultz. "Limitless Worker Surveillance." *California Law Review* 105, no. 3 (2017): 735–76. https://doi
.org/10.15779/z38br8mf94.

Ajunwa, Ifeoma, and Daniel Greene. "Platforms at Work: Automated Hiring
Platforms and other new Intermediaries in the Organization of Work."
In *Work and Labor in the Digital Age,* edited by Steven P. Vallas and Anne
Kovalainen, 66–91. Bingley, U.K.: Emerald, 2019.

"Albemarle (NYSE:ALB) Could Be Targeting These Nevada Lithium Juniors."
SmallCapPower, September 9, 2016. https://smallcappower.com/top
-stories/albemarle-nysealb-targeting-nevada-lithium-juniors/.

Alden, William. "Inside Palantir, Silicon Valley's Most Secretive Company."
Buzzfeed News, May 6, 2016. https://www.buzzfeednews.com/article
/williamalden/inside-palantir-silicon-valleys-most-secretive-company.

Aleksander, Igor, ed. *Artificial Vision for Robots.* Boston: Springer US, 1983.

"Amazon.Com Market Cap | AMZN." YCharts. https://ycharts.com/compa
nies/AMZN/market_cap.

"Amazon Rekognition Improves Face Analysis." Amazon Web Services, August 12, 2019. https://aws.amazon.com/about-aws/whats-new/2019/08
/amazon-rekognition-improves-face-analysis/.

"Amazon Rekognition—Video and Image—AWS." Amazon Web Services.
https://aws.amazon.com/rekognition/.

Ananny, Mike, and Kate Crawford. "Seeing without Knowing: Limitations
of the Transparency Ideal and Its Application to Algorithmic Accountability." *New Media and Society* 20, no. 3 (2018): 973–89. https://doi.org
/10.1177/1461444816676645.

Anderson, Warwick. *The Collectors of Lost Souls: Turning Kuru Scientists into
Whitemen.* Updated ed. Baltimore: Johns Hopkins University Press, 2019.

Andrae, Anders A. E., and Tomas Edler. "On Global Electricity Usage of
Communication Technology: Trends to 2030." *Challenges* 6, no. 1 (2015):
117–57. https://www.doi.org/10.3390/challe6010117.

Andrejevic, Mark. *Automated Media.* New York: Routledge, 2020.

Angwin, Julia, et al. "Dozens of Companies Are Using Facebook to Exclude
Older Workers from Job Ads." ProPublica, December 20, 2017. https://
www.propublica.org/article/facebook-ads-age-discrimination-targeting.

Angwin, Julia, et al. "Machine Bias." *ProPublica,* May 23, 2016. https://www
.propublica.org/article/machine-bias-risk-assessments-in-criminal-sen
tencing.

Anzilotti, Eillie. "Emails Show That ICE Uses Palantir Technology to Detain
Undocumented Immigrants," *FastCompany* (blog), July 16, 2019. https://

www.fastcompany.com/90377603/ice-uses-palantir-tech-to-detain
-immigrants-wnyc-report.

Apelbaum, Yaacov. "One Thousand and One Nights and Ilhan Omar's Biographical Engineering." *The Illustrated Primer* (blog), August 13, 2019.
https://apelbaum.wordpress.com/2019/08/13/one-thousand-and-one
-nights-and-ilhan-omars-biographical-engineering/.

Apple. "Apple Commits to Be 100 Percent Carbon Neutral for Its Supply
Chain and Products by 2030," July 21, 2020. https://www.apple.com/au
/newsroom/2020/07/apple-commits-to-be-100-percent-carbon-neutral
-for-its-supply-chain-and-products-by-2030/.

Apple. *Supplier Responsibility: 2018 Progress Report.* Cupertino, Calif.: Apple,
n.d. https://www.apple.com/supplier-responsibility/pdf/Apple_SR_2018
_Progress_Report.pdf.

Arboleda, Martin. *Planetary Mine: Territories of Extraction under Late Capitalism.* London: Verso, 2020.

Aristotle. *The Categories: On Interpretation.* Translated by Harold Percy
Cooke and Hugh Tredennick. Loeb Classical Library 325. Cambridge,
Mass.: Harvard University Press, 1938.

Aslam, Salman. "Facebook by the Numbers (2019): Stats, Demographics &
Fun Facts." Omnicore, January 6, 2020. https://www.omnicoreagency
.com/facebook-statistics/.

Ayogu, Melvin, and Zenia Lewis. "Conflict Minerals: An Assessment of the
Dodd-Frank Act." Brookings Institution, October 3, 2011. https://www
.brookings.edu/opinions/conflict-minerals-an-assessment-of-the-dodd
-frank-act/.

Aytes, Ayhan. "Return of the Crowds: Mechanical Turk and Neoliberal States
of Exception." In *Digital Labor: The Internet as Playground and Factory,*
edited by Trebor Scholz. New York: Routledge, 2013.

Babbage, Charles. *On the Economy of Machinery and Manufactures* [1832].
Cambridge: Cambridge University Press, 2010.

Babich, Babette E. *Nietzsche's Philosophy of Science: Reflecting Science on the
Ground of Art and Life.* Albany: State University of New York Press, 1994.

Bailey, F. G. "Dimensions of Rhetoric in Conditions of Uncertainty." In *Politically Speaking: Cross-Cultural Studies of Rhetoric,* edited by Robert
Paine, 25–38. Philadelphia: ISHI Press, 1981.

Baker, Janet M., et al. "Research Developments and Directions in Speech
Recognition and Understanding, Part 1." *IEEE,* April 2009. https://dspace
.mit.edu/handle/1721.1/51891.

Bangstad, Sindre, et al. "Thoughts on the Planetary: An Interview with
Achille Mbembé." *New Frame,* September 5, 2019. https://www.newframe
.com/thoughts-on-the-planetary-an-interview-with-achille-mbembe/.

Barrett, Lisa Feldman. "Are Emotions Natural Kinds?" *Perspectives on Psychological Science* 1, no. 1 (2006): 28–58. https://doi.org/10.1111/j.1745-6916.2006.00003.x.

Barrett, Lisa Feldman, et al. "Emotional Expressions Reconsidered: Challenges to Inferring Emotion from Human Facial Movements." *Psychological Science in the Public Interest* 20, no. 1 (2019): 1–68. https://doi.org/10.1177/1529100619832930.

"Bayan Obo Deposit, . . . Inner Mongolia, China." Mindat.org. https://www.mindat.org/loc-720.html.

Bayer, Ronald. *Homosexuality and American Psychiatry: The Politics of Diagnosis.* Princeton, N.J.: Princeton University Press, 1987.

Bechmann, Anja, and Geoffrey C. Bowker. "Unsupervised by Any Other Name: Hidden Layers of Knowledge Production in Artificial Intelligence on Social Media." *Big Data and Society* 6, no. 1 (2019): 205395171881956. https://doi.org/10.1177/2053951718819569.

Beck, Julie. "Hard Feelings: Science's Struggle to Define Emotions." *Atlantic,* February 24, 2015. https://www.theatlantic.com/health/archive/2015/02/hard-feelings-sciences-struggle-to-define-emotions/385711/.

Behrmann, Elisabeth, Jack Farchy, and Sam Dodge. "Hype Meets Reality as Electric Car Dreams Run into Metal Crunch." *Bloomberg,* January 11, 2018. https://www.bloomberg.com/graphics/2018-cobalt-batteries/.

Belkhir, L., and A. Elmeligi. "Assessing ICT Global Emissions Footprint: Trends to 2040 and Recommendations." *Journal of Cleaner Production* 177 (2018): 448–63.

Benjamin, Ruha. *Race after Technology: Abolitionist Tools for the New Jim Code.* Cambridge: Polity, 2019.

Benson, Kristina. "'Kill 'Em and Sort It Out Later': Signature Drone Strikes and International Humanitarian Law." *Pacific McGeorge Global Business and Development Law Journal* 27, no. 1 (2014): 17–51. https://www.mcgeorge.edu/documents/Publications/02_Benson_27_1.pdf.

Benthall, Sebastian, and Bruce D. Haynes. "Racial Categories in Machine Learning." In *FAT* '19: Proceedings of the Conference on Fairness, Accountability, and Transparency,* 289–98. New York: ACM Press, 2019. https://dl.acm.org/doi/10.1145/3287560.3287575.

Berg, Janine, et al. *Digital Labour Platforms and the Future of Work: Towards Decent Work in the Online World.* Geneva: International Labor Organization, 2018. https://www.ilo.org/wcmsp5/groups/public/---dgreports/---dcomm/---publ/documents/publication/wcms_645337.pdf.

Bergen, Mark. "Pentagon Drone Program Is Using Google AI." *Bloomberg,* March 6, 2018. https://www.bloomberg.com/news/articles/2018-03-06/google-ai-used-by-pentagon-drone-program-in-rare-military-pilot.

Berman, Sanford. *Prejudices and Antipathies: A Tract on the LC Subject Heads concerning People.* Metuchen, N.J.: Scarecrow Press, 1971.

Bezos, Jeff. *Going to Space to Benefit Earth.* Video, May 9, 2019. https://www.youtube.com/watch?v=GQ98hGUe6FM&.

Biddle, Wayne. *Dark Side of the Moon: Wernher von Braun, the Third Reich, and the Space Race.* New York: W. W. Norton, 2012.

Black, Edwin. *IBM and the Holocaust: The Strategic Alliance between Nazi Germany and America's Most Powerful Corporation.* Expanded ed. Washington, D.C.: Dialog Press, 2012.

Bledsoe, W. W. "The Model Method in Facial Recognition." Technical report, PRI 15. Palo Alto, Calif.: Panoramic Research, 1964.

Bloomfield, Anne B. "A History of the California Historical Society's New Mission Street Neighborhood." *California History* 74, no. 4 (1995–96): 372–93.

Blue, Violet. "Facebook Patents Tech to Determine Social Class." *Engadget,* February 9, 2018. https://www.engadget.com/2018-02-09-facebook-patents-tech-to-determine-social-class.html.

Blue Origin's Mission. Blue Origin. Video, February 1, 2019. https://www.youtube.com/watch?v=1YOL89kY8Og.

Bond, Charles F., Jr. "Commentary: A Few Can Catch a Liar, Sometimes: Comments on Ekman and O'Sullivan (1991), as Well as Ekman, O'Sullivan, and Frank (1999)." *Applied Cognitive Psychology* 22, no. 9 (2008): 1298–1300. https://doi.org/10.1002/acp.1475.

Borges, Jorge Luis. *Collected Fictions.* Translated by Andrew Hurley. New York: Penguin Books, 1998.

———. "John Wilkins' Analytical Language." In *Borges: Selected Non-Fictions,* edited by Eliot Weinberger. New York: Penguin Books, 2000.

———. *The Library of Babel.* Translated by Andrew Hurley. Boston: David R. Godine, 2000.

Bostrom, Nick. *Superintelligence: Paths, Dangers, Strategies.* Oxford: Oxford University Press, 2014.

Bouche, Teryn, and Laura Rivard. "America's Hidden History: The Eugenics Movement." Scitable, September 18, 2014. https://www.nature.com/scitable/forums/genetics-generation/america-s-hidden-history-the-eugenics-movement-123919444/.

Bowker, Geoffrey C. *Memory Practices in the Sciences.* Cambridge, Mass.: MIT Press, 2005.

Bowker, Geoffrey C., and Susan Leigh Star. *Sorting Things Out: Classification and Its Consequences.* Cambridge, Mass.: MIT Press, 1999.

Bratton, Benjamin H. *The Stack: On Software and Sovereignty.* Cambridge, Mass.: MIT Press, 2015.

Braverman, Harry. *Labor and Monopoly Capital: The Degradation of Work in the Twentieth Century.* 25th anniversary ed. New York: Monthly Review Press, 1998.

Brayne, Sarah. "Big Data Surveillance: The Case of Policing." *American Sociological Review* 82, no. 5 (2017): 977–1008. https://doi.org/10.1177/0003122417725865.

Brechin, Gray. *Imperial San Francisco: Urban Power, Earthly Ruin.* Berkeley: University of California Press, 2007.

Brewer, Eric. "Spanner, TrueTime and the CAP Theorem." Infrastructure: Google, February 14, 2017. https://storage.googleapis.com/pub-tools-pub lic-publication-data/pdf/45855.pdf.

Bridle, James. "Something Is Wrong on the Internet." *Medium* (blog), November 6, 2017. https://medium.com/@jamesbridle/something-is-wrong -on-the-internet-c39c471271d2.

Broussard, Meredith. *Artificial Unintelligence: How Computers Misunderstand the World.* Cambridge, Mass.: MIT Press, 2018.

Brown, Harold. *Department of Defense Annual Report: Fiscal Year 1982.* Report AD-A-096066/6. Washington, D.C., January 19, 1982. https://history .defense.gov/Portals/70/Documents/annual_reports/1982_DoD_AR.pdf ?ver=2014-06-24-150904-113.

Brown, Peter, and Robert Mercer. "Oh, Yes, Everything's Right on Schedule, Fred." Lecture, Twenty Years of Bitext Workshop, Empirical Methods in Natural Language Processing Conference, Seattle, Wash., October 2013. http://cs.jhu.edu/~post/bitext.

Browne, Simone. *Dark Matters: On the Surveillance of Blackness.* Durham, N.C.: Duke University Press, 2015.

———. "Digital Epidermalization: Race, Identity and Biometrics." *Critical Sociology* 36, no. 1 (January 2010): 131–50.

Brustein, Joshua, and Mark Bergen. "Google Wants to Do Business with the Military—Many of Its Employees Don't." *Bloomberg News,* November 21, 2019. https://www.bloomberg.com/features/2019-google-military-con tract-dilemma/.

Bullis, Kevin. "Lithium-Ion Battery." *MIT Technology Review,* June 19, 2012. https://www.technologyreview.com/s/428155/lithium-ion-battery/.

Buolamwini, Joy, and Timnit Gebru. "Gender Shades: Intersectional Accuracy Disparities in Commercial Gender Classification." *Proceedings of the First Conference on Fairness, Accountability and Transparency, PLMR* 81 (2018): 77–91. http://proceedings.mlr.press/v81/buolamwini18a.html.

Burke, Jason. "Congo Violence Fuels Fears of Return to 90s Bloodbath." *Guardian,* June 30, 2017. https://www.theguardian.com/world/2017/jun/30 /congo-violence-fuels-fears-of-return-to-90s-bloodbath.

Bush, Vannevar. "As We May Think." *Atlantic,* July 1945. https://www.the atlantic.com/magazine/archive/1945/07/as-we-may-think/303881/.

Business Council for Sustainable Energy. "2019 Sustainable Energy in America Factbook." BCSE, February 11, 2019. https://www.bcse.org/wp -content/uploads/2019-Sustainable-Energy-in-America-Factbook.pdf.

Byford, Sam. "Apple Buys Emotient, a Company That Uses AI to Read Emotions." *The Verge,* January 7, 2016. https://www.theverge.com/2016/1/7/10 731232/apple-emotient-ai-startup-acquisition.

"The CalGang Criminal Intelligence System." Sacramento: California State Auditor, Report 2015-130, August 2016. https://www.auditor.ca.gov/pdfs /reports/2015-130.pdf.

Calo, Ryan, and Danielle Citron. "The Automated Administrative State: A Crisis of Legitimacy" (March 9, 2020). *Emory Law Journal* (forthcoming). Available at SSRN: https://ssrn.com/abstract=3553590.

Cameron, Dell, and Kate Conger. "Google Is Helping the Pentagon Build AI for Drones." *Gizmodo,* March 6, 2018. https://gizmodo.com/google -is-helping-the-pentagon-build-ai-for-drones-1823464533.

Campolo, Alexander, and Kate Crawford. "Enchanted Determinism: Power without Responsibility in Artificial Intelligence." *Engaging Science, Technology, and Society* 6 (2020): 1–19. https://doi.org/10.17351/ests2020.277.

Canales, Jimena. *A Tenth of a Second: A History.* Chicago: University of Chicago Press, 2010.

Carey, James W. "Technology and Ideology: The Case of the Telegraph." *Prospects* 8 (1983): 303–25. https://doi.org/10.1017/S0361233300003793.

Carlisle, Nate. "NSA Utah Data Center Using More Water." *Salt Lake Tribune,* February 2, 2015. https://archive.sltrib.com/article.php?id=2118801 &itype=CMSID.

———. "Shutting Off NSA's Water Gains Support in Utah Legislature." *Salt Lake Tribune,* November 20, 2014. https://archive.sltrib.com/article .php?id=1845843&itype=CMSID.

Carter, Ash. *"Remarks on 'the Path to an Innovative Future for Defense' (CSIS Third Offset Strategy Conference)."* Washington, D.C.: U.S. Department of Defense, October 28, 2016. https://www.defense.gov/Newsroom/Speeches /Speech/Article/990315/remarks-on-the-path-to-an-innovative-future -for-defense-csis-third-offset-strat/.

Cave, Stephen, and Seán S. ÓhÉigeartaigh. "An AI Race for Strategic Advantage: Rhetoric and Risks." In *Proceedings of the 2018 AAAI/ACM Conference on AI, Ethics, and Society,* 36–40. https://dl.acm.org/doi/10.1145 /3278721.3278780.

Center for Land Use Interpretation, "Figure 2 Ranch, Texas," http://www.clui .org/ludb/site/figure-2-ranch.

Cetina, Karin Knorr. *Epistemic Cultures: How the Sciences Make Knowledge.* Cambridge, Mass.: Harvard University Press, 1999.

Champs, Emmanuelle de. "The Place of Jeremy Bentham's Theory of Fictions in Eighteenth-Century Linguistic Thought." *Journal of Bentham Studies* 2 (1999). https://doi.org/10.14324/111.2045-757X.011.

"Chinese Lithium Giant Agrees to Three-Year Pact to Supply Tesla." Bloomberg, September 21, 2018. https://www.industryweek.com/leadership/art icle/22026386/chinese-lithium-giant-agrees-to-threeyear-pact-to-sup ply-tesla.

Chinoy, Sahil. "Opinion: The Racist History behind Facial Recognition." *New York Times,* July 10, 2019. https://www.nytimes.com/2019/07/10 /opinion/facial-recognition-race.html.

Chun, Wendy Hui Kyong. *Control and Freedom: Power and Paranoia in the Age of Fiber Optics,* Cambridge, Mass: MIT Press, 2005.

Citton, Yves. *The Ecology of Attention.* Cambridge: Polity, 2017.

Clarac, François, Jean Massion, and Allan M. Smith. "Duchenne, Charcot and Babinski, Three Neurologists of La Salpetrière Hospital, and Their Contribution to Concepts of the Central Organization of Motor Synergy." *Journal of Physiology–Paris* 103, no. 6 (2009): 361–76. https://doi .org/10.1016/j.jphysparis.2009.09.001.

Clark, Nicola, and Simon Wallis. "Flamingos, Salt Lakes and Volcanoes: Hunting for Evidence of Past Climate Change on the High Altiplano of Bolivia." *Geology Today* 33, no. 3 (2017): 101–7. https://doi.org/10.1111/gto .12186.

Clauss, Sidonie. "John Wilkins' Essay toward a Real Character: Its Place in the Seventeenth-Century Episteme." *Journal of the History of Ideas* 43, no. 4 (1982): 531–53. https://doi.org/10.2307/2709342.

"'Clever Hans' Again: Expert Commission Decides That the Horse Actually Reasons.'" *New York Times,* October 2, 1904. https://timesmachine .nytimes.com/timesmachine/1904/10/02/120289067.pdf.

Cochran, Susan D., et al. "Proposed Declassification of Disease Categories Related to Sexual Orientation in the International Statistical Classification of Diseases and Related Health Problems (ICD-11)." *Bulletin of the World Health Organization* 92, no. 9 (2014): 672–79. https://doi.org/10 .2471/BLT.14.135541.

Cohen, Julie E. *Between Truth and Power: The Legal Constructions of Informational Capitalism.* New York: Oxford University Press, 2019.

Cole, David. "'We Kill People Based on Metadata.'" *New York Review of Books,* May 10, 2014. https://www.nybooks.com/daily/2014/05/10/we-kill -people-based-metadata/.

Colligan, Colette, and Margaret Linley, eds. *Media, Technology, and Litera-*

ture in the Nineteenth Century: Image, Sound, Touch. Burlington, VT: Ashgate, 2011.

"Colonized by Data: The Costs of Connection with Nick Couldry and Ulises Mejías." Book talk, September 19, 2019, Berkman Klein Center for Internet and Society at Harvard University. https://cyber.harvard.edu/events /colonized-data-costs-connection-nick-couldry-and-ulises-mejias.

"Congo's Bloody Coltan." Pulitzer Center on Crisis Reporting, January 6, 2011. https://pulitzercenter.org/reporting/congos-bloody-coltan.

Connolly, William E. *Climate Machines, Fascist Drives, and Truth.* Durham, N.C.: Duke University Press, 2019.

Connor, Neil. "Chinese School Uses Facial Recognition to Monitor Student Attention in Class." *Telegraph,* May 17, 2018. https://www.telegraph.co.uk /news/2018/05/17/chinese-school-uses-facial-recognition-monitor -student-attention/.

"Containers Lost at Sea—2017 Update." World Shipping Council, July 10, 2017. http://www.worldshipping.org/industry-issues/safety/Containers _Lost_at_Sea_-_2017_Update_FINAL_July_10.pdf.

Cook, Gary, et al. *Clicking Clean: Who Is Winning the Race to Build a Green Internet?* Washington, D.C.: Greenpeace, 2017. http://www.clickclean .org/international/en/.

Cook, James. "Amazon Patents New Alexa Feature That Knows When You're Ill and Offers You Medicine." *Telegraph,* October 9, 2018. https://www .telegraph.co.uk/technology/2018/10/09/amazon-patents-new-alexa -feature-knows-offers-medicine/.

Coole, Diana, and Samantha Frost, eds. *New Materialisms: Ontology, Agency, and Politics.* Durham, N.C.: Duke University Press, 2012.

Cooper, Carolyn C. "The Portsmouth System of Manufacture." *Technology and Culture* 25, no. 2 (1984): 182–225. https://doi.org/10.2307/3104712.

Corbett, James C., et al. "Spanner: Google's Globally-Distributed Database." *Proceedings of OSDI 2012* (2012): 14.

Costanza-Chock, Sasha. *Design Justice: Community-Led Practices to Build the Worlds We Need.* Cambridge, Mass.: MIT Press, 2020.

Couldry, Nick, and Ulises A. Mejías. *The Costs of Connection: How Data Is Colonizing Human Life and Appropriating It for Capitalism.* Stanford, Calif.: Stanford University Press, 2019.

———. "Data Colonialism: Rethinking Big Data's Relation to the Contemporary Subject." *Television and New Media* 20, no. 4 (2019): 336–49. https://doi.org/10.1177/1527476418796632.

"Counterpoints: An Atlas of Displacement and Resistance." *Anti-Eviction Mapping Project* (blog), September 3, 2020. https://antievictionmap.com /blog/2020/9/3/counterpoints-an-atlas-of-displacement-and-resistance.

Courtine, Jean-Jacques, and Claudine Haroche. *Histoire du visage: Exprimer et taire ses émotions (du XVIe siècle au début du XIXe siècle).* Paris: Payot et Rivages, 2007.

Cowen, Alan, et al. "Mapping the Passions: Toward a High-Dimensional Taxonomy of Emotional Experience and Expression." *Psychological Science in the Public Interest* 20, no. 1 (2019): 61–90. https://doi.org/10.1177/1529100619850176.

Crawford, Kate. "Halt the Use of Facial-Recognition Technology until It Is Regulated." *Nature* 572 (2019): 565. https://doi.org/10.1038/d41586-019-02514-7.

Crawford, Kate, and Vladan Joler. "Anatomy of an AI System." Anatomy of an AI System, 2018. http://www.anatomyof.ai.

Crawford, Kate, and Jason Schultz. "AI Systems as State Actors." *Columbia Law Review* 119, no. 7 (2019). https://columbialawreview.org/content/ai-systems-as-state-actors/.

———. "Big Data and Due Process: Toward a Framework to Redress Predictive Privacy Harms." *Boston College Law Review* 55, no. 1 (2014). https://lawdigitalcommons.bc.edu/bclr/vol55/iss1/4.

Crawford, Kate, et al. *AI Now 2019 Report.* New York: AI Now Institute, 2019. https://ainowinstitute.org/AI_Now_2019_Report.html.

Crevier, Daniel. *AI: The Tumultuous History of the Search for Artificial Intelligence.* New York: Basic Books, 1993.

Crosman, Penny. "Is AI a Threat to Fair Lending?" *American Banker,* September 7, 2017. https://www.americanbanker.com/news/is-artificial-intelligence-a-threat-to-fair-lending.

Currier, Cora, Glenn Greenwald, and Andrew Fishman. "U.S. Government Designated Prominent Al Jazeera Journalist as 'Member of Al Qaeda.'" *The Intercept* (blog), May 8, 2015. https://theintercept.com/2015/05/08/u-s-government-designated-prominent-al-jazeera-journalist-al-qaeda-member-put-watch-list/.

Curry, Steven, et al. "NIST Special Database 32: Multiple Encounter Dataset I (MEDS-I)." National Institute of Standards and Technology, NISTIR 7679, December 2009. https://nvlpubs.nist.gov/nistpubs/Legacy/IR/nistir7679.pdf.

Cuthbertson, Anthony. "Elon Musk Really Wants to 'Nuke Mars.'" *Independent,* August 19, 2019. https://www.independent.co.uk/life-style/gadgets-and-tech/news/elon-musk-mars-nuke-spacex-t-shirt-nuclear-weapons-space-a9069141.html.

Danowski, Déborah, and Eduardo Batalha Viveiros de Castro. *The Ends of the World.* Translated by Rodrigo Guimaraes Nunes. Malden, Mass.: Polity, 2017.

Danziger, Shai, Jonathan Levav, and Liora Avnaim-Pesso. "Extraneous Factors in Judicial Decisions." *Proceedings of the National Academy of Sciences* 108, no. 17 (2011): 6889–92. https://doi.org/10.1073/pnas.1018033108.

Darwin, Charles. *The Expression of the Emotions in Man and Animals,* edited by Joe Cain and Sharon Messenger. London: Penguin, 2009.

Dastin, Jeffrey. "Amazon Scraps Secret AI Recruiting Tool That Showed Bias against Women." *Reuters,* October 10, 2018. https://www.reuters.com/arti cle/us-amazon-com-jobs-automation-insight-idUSKCN1MK08G.

Daston, Lorraine. "Cloud Physiognomy." *Representations* 135, no. 1 (2016): 45–71. https://doi.org/10.1525/rep.2016.135.1.45.

Daston, Lorraine, and Peter Galison. *Objectivity.* Paperback ed. New York: Zone Books, 2010.

Davies, Kate, and Liam Young. *Tales from the Dark Side of the City: The Breastmilk of the Volcano, Bolivia and the Atacama Desert Expedition.* London: Unknown Fields, 2016.

Davis, F. James. *Who Is Black? One Nation's Definition.* 10th anniversary ed. University Park: Pennsylvania State University Press, 2001.

Davis, Monte. "Gerard K. O'Neill on Space Colonies." *Omni Magazine,* October 12, 2017. https://omnimagazine.com/interview-gerard-k-oneill -space-colonies/.

Delaporte, François. *Anatomy of the Passions.* Translated by Susan Emanuel. Stanford, Calif.: Stanford University Press, 2008.

Demis Hassabis, DeepMind — Learning from First Principles — Artificial Intel- ligence NIPS2017. Video, December 9, 2017. https://www.youtube.com /watch?v=DXNqYSNvnjA&feature=emb_title.

Deng, Jia, et al. "ImageNet: A Large-Scale Hierarchical Image Database." In *2009 IEEE Conference on Computer Vision and Pattern Recognition,* 248–55. https://doi.org/10.1109/CVPR.2009.5206848.

Department of International Cooperation, Ministry of Science and Tech- nology. "Next Generation Artificial Intelligence Development Plan." *China Science and Technology Newsletter,* no. 17, September 15, 2017. http://fi.china-embassy.org/eng/kxjs/P020171025789108009001.pdf.

Deputy Secretary of Defense to Secretaries of the Military Departments et al., April 26, 2017. Memorandum: "Establishment of an Algorithmic Warfare Cross-Functional Team (Project Maven)." https://www.govexec .com/media/gbc/docs/pdfs_edit/establishment_of_the_awcft_project _maven.pdf.

Derrida, Jacques, and Eric Prenowitz. "Archive Fever: A Freudian Impres- sion." *Diacritics* 25, no. 2 (1995): 9. https://doi.org/10.2307/465144.

Dick, Philip K. "Autofac." *Galaxy Magazine,* November 1955. http://archive .org/details/galaxymagazine-1955-11.

Didi-Huberman, Georges. *Atlas, or the Anxious Gay Science: How to Carry the World on One's Back?* Chicago: University of Chicago Press, 2018.

Dietterich, Thomas, and Eun Bae Kong. "Machine Learning Bias, Statistical Bias, and Statistical Variance of Decision Tree Algorithms." Unpublished paper, Oregon State University, 1995. http://citeseerx.ist.psu.edu/view doc/summary?doi=10.1.1.38.2702.

D'Ignazio, Catherine, and Lauren F. Klein. *Data Feminism*. Cambridge, Mass.: MIT Press, 2020.

Dilanian, Ken. "US Special Operations Forces Are Clamoring to Use Software from Silicon Valley Company Palantir." *Business Insider,* March 26, 2015. https://www.businessinsider.com/us-special-operations-forces-are -clamoring-to-use-software-from-silicon-valley-company-palantir-2015-3.

Dobbe, Roel, and Meredith Whittaker. "AI and Climate Change: How They're Connected, and What We Can Do about It." *Medium* (blog), October 17, 2019. https://medium.com/@AINowInstitute/ai-and-climate-change -how-theyre-connected-and-what-we-can-do-about-it-6aa8d0f5b32c.

Domingos, Pedro. "A Few Useful Things to Know about Machine Learning." *Communications of the ACM* 55, no. 10 (2012): 78. https://doi.org/10.1145 /2347736.2347755.

Dooley, Ben, Eimi Yamamitsu, and Makiko Inoue. "Fukushima Nuclear Disaster Trial Ends with Acquittals of 3 Executives." *New York Times,* September 19, 2019. https://www.nytimes.com/2019/09/19/business/japan -tepco-fukushima-nuclear-acquitted.html.

Dougherty, Conor. "Google Photos Mistakenly Labels Black People 'Gorillas.'" *Bits Blog* (blog), July 1, 2015. https://bits.blogs.nytimes.com/2015/07/01 /google-photos-mistakenly-labels-black-people-gorillas/.

Douglass, Frederick. "West India Emancipation." Speech delivered at Canandaigua, N.Y., August 4, 1857. https://rbscp.lib.rochester.edu/4398.

Drescher, Jack. "Out of DSM: Depathologizing Homosexuality." *Behavioral Sciences* 5, no. 4 (2015): 565–75. https://doi.org/10.3390/bs5040565.

Dreyfus, Hubert L. *Alchemy and Artificial Intelligence*. Santa Monica, Calif.: RAND, 1965.

———. *What Computers Can't Do: A Critique of Artificial Reason*. New York: Harper and Row, 1972.

Dryer, Theodora. "Designing Certainty: The Rise of Algorithmic Computing in an Age of Anxiety 1920–1970." Ph.D. diss., University of California, San Diego, 2019.

Du, Lisa, and Ayaka Maki. "AI Cameras That Can Spot Shoplifters Even before They Steal." *Bloomberg,* March 4, 2019. https://www.bloomberg.com /news/articles/2019-03-04/the-ai-cameras-that-can-spot-shoplifters -even-before-they-steal.

Duchenne (de Boulogne), G.-B. *Mécanisme de la physionomie humaine ou Analyse électro-physiologique de l'expression des passions applicable à la pratique des arts plastiques.* 2nd ed. Paris: Librairie J.-B. Baillière et Fils, 1876.

Eco, Umberto. *The Infinity of Lists: An Illustrated Essay.* Translated by Alastair McEwen. New York: Rizzoli, 2009.

Edwards, Paul N. *The Closed World: Computers and the Politics of Discourse in Cold War America.* Cambridge, Mass.: MIT Press, 1996.

Edwards, Paul N., and Gabrielle Hecht. "History and the Technopolitics of Identity: The Case of Apartheid South Africa." *Journal of Southern African Studies* 36, no. 3 (2010): 619–39. https://doi.org/10.1080/03057070.2010.507568.

Eglash, Ron. "Broken Metaphor: The Master-Slave Analogy in Technical Literature." *Technology and Culture* 48, no. 2 (2007): 360–69. https://doi.org/10.1353/tech.2007.0066.

Ekman, Paul. "An Argument for Basic Emotions." *Cognition and Emotion* 6, no. 3–4 (1992): 169–200.

———. "Duchenne and Facial Expression of Emotion." In G.-B. Duchenne de Boulogne, *The Mechanism of Human Facial Expression*, 270–84. Edited and translated by R. A. Cuthbertson. Cambridge: Cambridge University Press, 1990.

———. *Emotions Revealed: Recognizing Faces and Feelings to Improve Communication and Emotional Life.* New York: Times Books, 2003.

———. "A Life's Pursuit." In *The Semiotic Web '86: An International Yearbook,* edited by Thomas A. Sebeok and Jean Umiker-Sebeok, 4–46. Berlin: Mouton de Gruyter, 1987.

———. *Nonverbal Messages: Cracking the Code: My Life's Pursuit.* San Francisco: PEG, 2016.

———. *Telling Lies: Clues to Deceit in the Marketplace, Politics, and Marriage.* 4th ed. New York: W. W. Norton, 2009.

———. "Universal Facial Expressions of Emotion." *California Mental Health Research Digest* 8, no. 4 (1970): 151–58.

———. "What Scientists Who Study Emotion Agree About." *Perspectives on Psychological Science* 11, no. 1 (2016): 81–88. https://doi.org/10.1177/1745691615596992.

Ekman, Paul, and Wallace V. Friesen. "Constants across Cultures in the Face and Emotion." *Journal of Personality and Social Psychology* 17, no. 2 (1971): 124–29. https://doi.org/10.1037/h0030377.

———. *Facial Action Coding System (FACS): A Technique for the Measurement of Facial Action.* Palo Alto, Calif.: Consulting Psychologists Press, 1978.

———. "Nonverbal Leakage and Clues to Deception." *Psychiatry* 31, no. 1 (1969): 88–106.

————. *Unmasking the Face*. Cambridge, Mass.: Malor Books, 2003.

Ekman, Paul, and Harriet Oster. "Facial Expressions of Emotion." *Annual Review of Psychology* 30 (1979): 527–54.

Ekman, Paul, and Maureen O'Sullivan. "Who Can Catch a Liar?" *American Psychologist* 46, no. 9 (1991): 913–20. https://doi.org/10.1037/0003-066X.46.9.913.

Ekman, Paul, Maureen O'Sullivan, and Mark G. Frank. "A Few Can Catch a Liar." *Psychological Science* 10, no. 3 (1999): 263–66. https://doi.org/10.1111/1467-9280.00147.

Ekman, Paul, and Erika L. Rosenberg, eds. *What the Face Reveals: Basic and Applied Studies of Spontaneous Expression Using the Facial Action Coding System (FACS)*. New York: Oxford University Press, 1997.

Ekman, Paul, E. Richard Sorenson, and Wallace V. Friesen. "Pan-Cultural Elements in Facial Displays of Emotion." *Science* 164 (1969): 86–88. https://doi.org/10.1126/science.164.3875.86.

Ekman, Paul, et al. "Universals and Cultural Differences in the Judgments of Facial Expressions of Emotion." *Journal of Personality and Social Psychology* 53, no. 4 (1987): 712–17.

Elfenbein, Hillary Anger, and Nalini Ambady. "On the Universality and Cultural Specificity of Emotion Recognition: A Meta-Analysis." *Psychological Bulletin* 128, no. 2 (2002): 203–35. https://doi.org/10.1037/0033-2909.128.2.203.

Elish, Madeline Clare, and danah boyd. "Situating Methods in the Magic of Big Data and AI." *Communication Monographs* 85, no. 1 (2018): 57–80. https://doi.org/10.1080/03637751.2017.1375130.

Ely, Chris. "The Life Expectancy of Electronics." Consumer Technology Association, September 16, 2014. https://www.cta.tech/News/Blog/Articles/2014/September/The-Life-Expectancy-of-Electronics.aspx.

"Emotion Detection and Recognition (EDR) Market Size to surpass 18%+ CAGR 2020 to 2027." *MarketWatch*, October 5, 2020. https://www.marketwatch.com/press-release/emotion-detection-and-recognition-edr-market-size-to-surpass-18-cagr-2020-to-2027-2020-10-05.

England, Rachel. "UK Police's Facial Recognition System Has an 81 Percent Error Rate." *Engadget*, July 4, 2019. https://www.engadget.com/2019/07/04/uk-met-facial-recognition-failure-rate/.

Ensmenger, Nathan. "Computation, Materiality, and the Global Environment." *IEEE Annals of the History of Computing* 35, no. 3 (2013): 80. https://www.doi.org/10.1109/MAHC.2013.33.

————. *The Computer Boys Take Over: Computers, Programmers, and the Politics of Technical Expertise*. Cambridge, Mass.: MIT Press, 2010.

Eschner, Kat. "Lie Detectors Don't Work as Advertised and They Never Did." *Smithsonian*, February 2, 2017. https://www.smithsonianmag.com/smart

-news/lie-detectors-dont-work-advertised-and-they-never-did-18096
1956/.

Estreicher, Sam, and Christopher Owens. "Labor Board Wrongly Rejects Employee Access to Company Email for Organizational Purposes." *Verdict,* February 19, 2020. https://verdict.justia.com/2020/02/19/labor-board-wrongly-rejects-employee-access-to-company-email-for-organizational-purposes.

Eubanks, Virginia. *Automating Inequality: How High-Tech Tools Profile, Police, and Punish the Poor.* New York: St. Martin's, 2017.

Ever AI. "Ever AI Leads All US Companies on NIST's Prestigious Facial Recognition Vendor Test." *GlobeNewswire,* November 27, 2018. http://www.globenewswire.com/news-release/2018/11/27/1657221/0/en/Ever-AI-Leads-All-US-Companies-on-NIST-s-Prestigious-Facial-Recognition-Vendor-Test.html.

Fabian, Ann. *The Skull Collectors: Race, Science, and America's Unburied Dead.* Chicago: University of Chicago Press, 2010.

"Face: An AI Service That Analyzes Faces in Images." Microsoft Azure. https://azure.microsoft.com/en-us/services/cognitive-services/face/.

Fadell, Anthony M., et al. Smart-home automation system that suggests or automatically implements selected household policies based on sensed observations. US10114351B2, filed March 5, 2015, and issued October 30, 2018.

Fang, Lee. "Defense Tech Startup Founded by Trump's Most Prominent Silicon Valley Supporters Wins Secretive Military AI Contract." *The Intercept* (blog), March 9, 2019. https://theintercept.com/2019/03/09/anduril-industries-project-maven-palmer-luckey/.

———. "Leaked Emails Show Google Expected Lucrative Military Drone AI Work to Grow Exponentially." *The Intercept* (blog), May 31, 2018. https://theintercept.com/2018/05/31/google-leaked-emails-drone-ai-pentagon-lucrative/.

"Federal Policy for the Protection of Human Subjects." *Federal Register,* September 8, 2015. https://www.federalregister.gov/documents/2015/09/08/2015-21756/federal-policy-for-the-protection-of-human-subjects.

Federici, Silvia. *Wages against Housework.* 6th ed. London: Power of Women Collective and Falling Walls Press, 1975.

Fellbaum, Christiane, ed. *WordNet: An Electronic Lexical Database.* Cambridge, Mass.: MIT Press, 1998.

Fernández-Dols, José-Miguel, and James A. Russell, eds. *The Science of Facial Expression.* New York: Oxford University Press, 2017.

Feuer, William. "Palantir CEO Alex Karp Defends His Company's Relationship with Government Agencies." *CNBC,* January 23, 2020. https://www.cnbc.com/2020/01/23/palantir-ceo-alex-karp-defends-his-companys-work-for-the-government.html.

"Five Eyes Intelligence Oversight and Review Council." U.S. Office of the Director of National Intelligence. https://www.dni.gov/index.php/who-we-are/organizations/enterprise-capacity/chco/chco-related-menus/chco-related-links/recruitment-and-outreach/217-about/organization/icig-pages/2660-icig-fiorc.

Foer, Franklin. "Jeff Bezos's Master Plan." *Atlantic,* November 2019. https://www.theatlantic.com/magazine/archive/2019/11/what-jeff-bezos-wants/598363/.

Foreman, Judy. "A Conversation with: Paul Ekman; The 43 Facial Muscles That Reveal Even the Most Fleeting Emotions." *New York Times,* August 5, 2003. https://www.nytimes.com/2003/08/05/health/conversation-with-paul-ekman-43-facial-muscles-that-reveal-even-most-fleeting.html.

Forsythe, Diana E. "Engineering Knowledge: The Construction of Knowledge in Artificial Intelligence." *Social Studies of Science* 23, no. 3 (1993): 445–77. https://doi.org/10.1177/0306312793023003002.

Fortunati, Leopoldina. "Robotization and the Domestic Sphere." *New Media and Society* 20, no. 8 (2018): 2673–90. https://doi.org/10.1177/1461444481 7729366.

Foucault, Michel. *Discipline and Punish: The Birth of the Prison.* 2nd ed. New York: Vintage Books, 1995.

Founds, Andrew P., et al. "NIST Special Database 32: Multiple Encounter Dataset II (MEDS-II)." National Institute of Standards and Technology, NISTIR 7807, February 2011. https://tsapps.nist.gov/publication/get_pdf.cfm?pub_id=908383.

Fourcade, Marion, and Kieran Healy. "Seeing Like a Market." *Socio-Economic Review* 15, no. 1 (2016): 9–29. https://doi.org/10.1093/ser/mww033.

Franceschi-Bicchierai, Lorenzo. "Redditor Cracks Anonymous Data Trove to Pinpoint Muslim Cab Drivers." *Mashable,* January 28, 2015. https://mashable.com/2015/01/28/redditor-muslim-cab-drivers/.

Franklin, Ursula M. *The Real World of Technology.* Rev. ed. Toronto, Ont.: House of Anansi Press, 2004.

Franklin, Ursula M., and Michelle Swenarchuk. *The Ursula Franklin Reader: Pacifism as a Map.* Toronto, Ont.: Between the Lines, 2006.

French, Martin A., and Simone A. Browne. "Surveillance as Social Regulation: Profiles and Profiling Technology." In *Criminalization, Representation, Regulation: Thinking Differently about Crime,* edited by Deborah R. Brock, Amanda Glasbeek, and Carmela Murdocca, 251–84. North York, Ont.: University of Toronto Press, 2014.

Fridlund, Alan. "A Behavioral Ecology View of Facial Displays, 25 Years Later." *Emotion Researcher,* August 2015. https://emotionresearcher.com/the-behavioral-ecology-view-of-facial-displays-25-years-later/.

Fussell, Sidney. "The Next Data Mine Is Your Bedroom." *Atlantic,* Novem-

ber 17, 2018. https://www.theatlantic.com/technology/archive/2018/11
/google-patent-bedroom-privacy-smart-home/576022/.

Galison, Peter. *Einstein's Clocks, Poincaré's Maps: Empires of Time.* New York:
W. W. Norton, 2003.

———. "The Ontology of the Enemy: Norbert Wiener and the Cybernetic
Vision." *Critical Inquiry* 21, no. 1 (1994): 228–66.

———. "Removing Knowledge." *Critical Inquiry* 31, no. 1 (2004): 229–43.
https://doi.org/10.1086/427309.

Garris, Michael D., and Charles L. Wilson. "NIST Biometrics Evaluations
and Developments." National Institute of Standards and Technology,
NISTIR 7204, February 2005. https://www.govinfo.gov/content/pkg
/GOVPUB-C13-1ba4778e3b87bdd6ce660349317d3263/pdf/GOVPUB
-C13-1ba4778e3b87bdd6ce660349317d3263.pdf.

Gates, Dominic. "Bezos's Blue Origin Seeks Tax Incentives to Build Rocket
Engines Here." *Seattle Times,* January 14, 2016. https://www.seattletimes
.com/business/boeing-aerospace/bezoss-blue-origin-seeks-tax-incen
tives-to-build-rocket-engines-here/.

Gebru, Timnit, et al. "Datasheets for Datasets." *ArXiv:1803.09010 [Cs],* March
23, 2018. http://arxiv.org/abs/1803.09010.

———. "Fine-Grained Car Detection for Visual Census Estimation." In
*Proceedings of the Thirty-First AAAI Conference on Artificial Intelligence,
AAAI '17,* 4502–8.

Gee, Alastair. "San Francisco or Mumbai? UN Envoy Encounters Homeless
Life in California." *Guardian,* January 22, 2018. https://www.theguardian
.com/us-news/2018/jan/22/un-rapporteur-homeless-san-francisco-cali
fornia.

Gellman, Barton, and Laura Poitras. "U.S., British Intelligence Mining Data
from Nine U.S. Internet Companies in Broad Secret Program." *Washington Post,* June 7, 2013. https://www.washingtonpost.com/investigations
/us-intelligence-mining-data-from-nine-us-internet-companies-in
-broad-secret-program/2013/06/06/3a0c0da8-cebf-11e2-8845-d970ccb0
4497_story.html.

Gendron, Maria, and Lisa Feldman Barrett. *Facing the Past.* Vol. 1. New York:
Oxford University Press, 2017.

George, Rose. *Ninety Percent of Everything: Inside Shipping, the Invisible Industry That Puts Clothes on Your Back, Gas in Your Car, and Food on Your
Plate.* New York: Metropolitan Books, 2013.

Gershgorn, Dave. "The Data That Transformed AI Research—and Possibly
the World." *Quartz,* July 26, 2017. https://qz.com/1034972/the-data-that
-changed-the-direction-of-ai-research-and-possibly-the-world/.

Ghaffary, Shirin. "More Than 1,000 Google Employees Signed a Letter Demanding the Company Reduce Its Carbon Emissions." *Recode,* Novem-

ber 4, 2019. https://www.vox.com/recode/2019/11/4/20948200/google-em
ployees-letter-demand-climate-change-fossil-fuels-carbon-emissions.

Gill, Karamjit S. *Artificial Intelligence for Society.* New York: John Wiley and Sons, 1986.

Gillespie, Tarleton. *Custodians of the Internet: Platforms, Content Modera-tion, and the Hidden Decisions That Shape Social Media.* New Haven: Yale University Press, 2018.

Gillespie, Tarleton, Pablo J. Boczkowski, and Kirsten A. Foot, eds. *Media Technologies: Essays on Communication, Materiality, and Society.* Cam-bridge. Mass.: MIT Press, 2014.

Gitelman, Lisa, ed. *"Raw Data" Is an Oxymoron.* Cambridge, Mass.: MIT Press, 2013.

Goeleven, Ellen, et al. "The Karolinska Directed Emotional Faces: A Valida-tion Study." *Cognition and Emotion* 22, no. 6 (2008): 1094–18. https://doi .org/10.1080/02699930701626582.

Goenka, Aakash, et al. Database systems and user interfaces for dynamic and interactive mobile image analysis and identification. US10339416B2, filed July 5, 2018, and issued July 2, 2019.

"Google Outrage at 'NSA Hacking.'" *BBC News,* October 31, 2013. https:// www.bbc.com/news/world-us-canada-24751821.

Gora, Walter, Ulrich Herzog, and Satish Tripathi. "Clock Synchronization on the Factory Floor (FMS)." *IEEE Transactions on Industrial Electronics* 35, no. 3 (1988): 372–80. https://doi.org/10.1109/41.3109.

Gould, Stephen Jay. *The Mismeasure of Man.* Rev. and expanded ed. New York: W. W. Norton, 1996.

Graeber, David. *The Utopia of Rules: On Technology, Stupidity, and the Secret Joys of Bureaucracy.* Brooklyn, N.Y.: Melville House, 2015.

Graham, John. "Lavater's Physiognomy in England." *Journal of the History of Ideas* 22, no. 4 (1961): 561. https://doi.org/10.2307/2708032.

Graham, Mark, and Håvard Haarstad. "Transparency and Development: Ethical Consumption through Web 2.0 and the Internet of Things." *Infor-mation Technologies and International Development* 7, no. 1 (2011): 1–18.

Gray, Mary L., and Siddharth Suri. *Ghost Work: How to Stop Silicon Valley from Building a New Global Underclass.* Boston: Houghton Mifflin Har-court, 2019.

———. "The Humans Working behind the AI Curtain." *Harvard Business Review,* January 9, 2017. https://hbr.org/2017/01/the-humans-working -behind-the-ai-curtain.

Gray, Richard T. *About Face: German Physiognomic Thought from Lavater to Auschwitz.* Detroit, Mich.: Wayne State University Press, 2004.

Green, Ben. *Smart Enough City: Taking Off Our Tech Goggles and Reclaiming the Future of Cities.* Cambridge, Mass.: MIT Press, 2019.

Greenberger, Martin, ed. *Management and the Computer of the Future.* New York: Wiley, 1962.

Greene, Tristan. "Science May Have Cured Biased AI." The Next Web, October 26, 2017. https://thenextweb.com/artificial-intelligence/2017/10/26/scientists-may-have-just-created-the-cure-for-biased-ai/.

Greenhouse, Steven. "McDonald's Workers File Wage Suits in 3 States." *New York Times,* March 13, 2014. https://www.nytimes.com/2014/03/14/business/mcdonalds-workers-in-three-states-file-suits-claiming-under payment.html.

Greenwald, Anthony G., and Linda Hamilton Krieger. "Implicit Bias: Scientific Foundations." *California Law Review* 94, no. 4 (2006): 945. https://doi.org/10.2307/20439056.

Gregg, Melissa. *Counterproductive: Time Management in the Knowledge Economy.* Durham, N.C.: Duke University Press, 2018.

"A Grey Goldmine: Recent Developments in Lithium Extraction in Bolivia and Alternative Energy Projects." Council on Hemispheric Affairs, November 17, 2009. http://www.coha.org/a-grey-goldmine-recent-develop ments-in-lithium-extraction-in-bolivia-and-alternative-energy-projects/.

Grigorieff, Paul. "The Mittelwerk/Mittelbau/Camp Dora." V2rocket.com. http://www.v2rocket.com/start/chapters/mittel.html.

Grother, Patrick, et al. "The 2017 IARPA Face Recognition Prize Challenge (FRPC)." National Institute of Standards and Technology, NISTIR 8197, November 2017. https://nvlpubs.nist.gov/nistpubs/ir/2017/NIST.IR.8197.pdf.

Grothoff, Christian, and J. M. Porup. "The NSA's SKYNET Program May Be Killing Thousands of Innocent People." Ars Technica, February 16, 2016. https://arstechnica.com/information-technology/2016/02/the-nsas-sky net-program-may-be-killing-thousands-of-innocent-people/.

Guendelsberger, Emily. *On the Clock: What Low-Wage Work Did to Me and How It Drives America Insane.* New York: Little, Brown, 2019.

Gurley, Lauren Kaori. "60 Amazon Workers Walked Out over Warehouse Working Conditions." *Vice* (blog), October 3, 2019. https://www.vice.com/en_us/article/pa7qny/60-amazon-workers-walked-out-over-ware house-working-conditions.

Hacking, Ian. "Kinds of People: Moving Targets." *Proceedings of the British Academy* 151 (2007): 285–318.

———. "Making Up People." *London Review of Books,* August 17, 2006, 23–26.

Hagendorff, Thilo. "The Ethics of AI Ethics: An Evaluation of Guidelines." *Minds and Machines* 30 (2020): 99–120. https://doi.org/10.1007/s11023-020-09517-8.

Haggerty, Kevin D., and Richard V. Ericson. "The Surveillant Assemblage."

British Journal of Sociology 51, no. 4 (2000): 605–22. https://doi.org/10
.1080/00071310020015280.

Hajjar, Lisa. "Lawfare and Armed Conflicts: A Comparative Analysis of Israeli and U.S. Targeted Killing Policies." In *Life in the Age of Drone Warfare*, edited by Lisa Parks and Caren Kaplan, 59–88. Durham, N.C.: Duke University Press, 2017.

Halsey III, Ashley. "House Member Questions $900 Million TSA 'SPOT' Screening Program." *Washington Post,* November 14, 2013. https://www .washingtonpost.com/local/trafficandcommuting/house-member-ques tions-900-million-tsa-spot-screening-program/2013/11/14/ad194cfe -4d5c-11e3-be6b-d3d28122e6d4_story.html.

Hao, Karen. "AI Is Sending People to Jail—and Getting It Wrong." *MIT Technology Review,* January 21, 2019. https://www.technologyreview.com/s /612775/algorithms-criminal-justice-ai/.

———. "The Technology behind OpenAI's Fiction-Writing, Fake-News-Spewing AI, Explained." *MIT Technology Review,* February 16, 2019. https://www.technologyreview.com/s/612975/ai-natural-language -processing-explained/.

———. "Three Charts Show How China's AI Industry Is Propped Up by Three Companies." *MIT Technology Review,* January 22, 2019. https:// www.technologyreview.com/s/612813/the-future-of-chinas-ai-industry -is-in-the-hands-of-just-three-companies/.

Haraway, Donna J. *Modest_Witness@Second_Millennium.FemaleMan_Meets_OncoMouse: Feminism and Technoscience.* New York: Routledge, 1997.

———. *Simians, Cyborgs, and Women: The Reinvention of Nature.* New York: Routledge, 1990.

———. *When Species Meet.* Minneapolis: University of Minnesota Press, 2008.

Hardt, Michael, and Antonio Negri. *Assembly.* New York: Oxford University Press, 2017.

Harrabin, Roger. "Google Says Its Carbon Footprint Is Now Zero." BBC News, September 14, 2020. https://www.bbc.com/news/technology-54141899.

Harvey, Adam R. "MegaPixels." MegaPixels. https://megapixels.cc/.

Harvey, Adam, and Jules LaPlace. "Brainwash Dataset." MegaPixels. https:// megapixels.cc/brainwash/.

Harwell, Drew. "A Face-Scanning Algorithm Increasingly Decides Whether You Deserve the Job." *Washington Post,* November 7, 2019. https://www .washingtonpost.com/technology/2019/10/22/ai-hiring-face-scanning -algorithm-increasingly-decides-whether-you-deserve-job/.

Haskins, Caroline. "Amazon Is Coaching Cops on How to Obtain Surveillance Footage without a Warrant." *Vice* (blog), August 5, 2019. https://

www.vice.com/en_us/article/43kga3/amazon-is-coaching-cops-on
-how-to-obtain-surveillance-footage-without-a-warrant.

———. "Amazon's Home Security Company Is Turning Everyone into Cops." *Vice* (blog), February 7, 2019. https://www.vice.com/en_us/article /qvyvzd/amazons-home-security-company-is-turning-everyone-into -cops.

———. "How Ring Transmits Fear to American Suburbs." *Vice* (blog), July 12, 2019. https://www.vice.com/en/article/ywaa57/how-ring-transmits -fear-to-american-suburbs.

Heaven, Douglas. "Why Faces Don't Always Tell the Truth about Feelings." *Nature*, February 26, 2020. https://www.nature.com/articles/d41586-020 -00507-5.

Heller, Nathan. "What the Enron Emails Say about Us." *New Yorker*, July 17, 2017. https://www.newyorker.com/magazine/2017/07/24/what-the -enron-e-mails-say-about-us.

Hernandez, Elizabeth. "CU Colorado Springs Students Secretly Photo-graphed for Government-Backed Facial-Recognition Research." *Denver Post*, May 27, 2019. https://www.denverpost.com/2019/05/27/cu-colorado -springs-facial-recognition-research/.

Heyn, Edward T. "Berlin's Wonderful Horse; He Can Do Almost Everything but Talk—How He Was Taught." *New York Times*, Sept. 4, 1904. https:// timesmachine.nytimes.com/timesmachine/1904/09/04/101396572.pdf.

Hicks, Mar. *Programmed Inequality: How Britain Discarded Women Tech-nologists and Lost Its Edge in Computing.* Cambridge, Mass.: MIT Press, 2017.

Hird, M. J. "Waste, Landfills, and an Environmental Ethics of Vulnerability." *Ethics and the Environment* 18, no. 1 (2013): 105–24. https://www.doi.org /10.2979/ethicsenviro.18.1.105.

Hodal, Kate. "Death Metal: Tin Mining in Indonesia." *Guardian*, November 23, 2012. https://www.theguardian.com/environment/2012/nov/23/tin -mining-indonesia-bangka.

Hoffmann, Anna Lauren. "Data Violence and How Bad Engineering Choices Can Damage Society." *Medium* (blog), April 30, 2018. https://medium .com/s/story/data-violence-and-how-bad-engineering-choices-can-dam age-society-39e44150e1d4.

Hoffower, Hillary. "We Did the Math to Calculate How Much Money Jeff Bezos Makes in a Year, Month, Week, Day, Hour, Minute, and Second." *Business Insider*, January 9, 2019. https://www.businessinsider.com/what -amazon-ceo-jeff-bezos-makes-every-day-hour-minute-2018-10.

Hoft, Joe. "Facial, Speech and Virtual Polygraph Analysis Shows Ilhan Omar Exhibits Many Indications of a Compulsive Fibber!!!" The Gateway Pun-dit, July 21, 2019. https://www.thegatewaypundit.com/2019/07/facial-spe

ech-and-virtual-polygraph-analysis-shows-ilhan-omar-exhibits-many
-indications-of-a-compulsive-fibber/.

Hogan, Mél. "Data Flows and Water Woes: The Utah Data Center." *Big Data and Society* (December 2015). https://www.doi.org/10.1177/2053951715592429.

Holmqvist, Caroline. *Policing Wars: On Military Intervention in the Twenty-First Century.* London: Palgrave Macmillan, 2014.

Horne, Emily, and Tim Maly. *The Inspection House: An Impertinent Field Guide to Modern Surveillance.* Toronto: Coach House Books, 2014.

Horowitz, Alexandra. "Why Brain Size Doesn't Correlate with Intelligence." *Smithsonian,* December 2013. https://www.smithsonianmag.com/science-nature/why-brain-size-doesnt-correlate-with-intelligence-180947627/.

House, Brian. "Synchronizing Uncertainty: Google's Spanner and Cartographic Time." In *Executing Practices,* edited by Helen Pritchard, Eric Snodgrass, and Magda Tyżlik-Carver, 117–26. London: Open Humanities Press, 2018.

"How Does a Lithium-Ion Battery Work?" Energy.gov, September 14, 2017. https://www.energy.gov/eere/articles/how-does-lithium-ion-battery-work.

Hu, Tung-Hui. *A Prehistory of the Cloud.* Cambridge, Mass.: MIT Press, 2015.

Huet, Ellen. "The Humans Hiding behind the Chatbots." *Bloomberg,* April 18, 2016. https://www.bloomberg.com/news/articles/2016-04-18/the-humans-hiding-behind-the-chatbots.

Hutson, Matthew. "Artificial Intelligence Could Identify Gang Crimes—and Ignite an Ethical Firestorm." *Science,* February 28, 2018. https://www.sciencemag.org/news/2018/02/artificial-intelligence-could-identify-gang-crimes-and-ignite-ethical-firestorm.

Hwang, Tim, and Karen Levy. "'The Cloud' and Other Dangerous Metaphors." *Atlantic,* January 20, 2015. https://www.theatlantic.com/technology/archive/2015/01/the-cloud-and-other-dangerous-metaphors/384518/.

"ImageNet Large Scale Visual Recognition Competition (ILSVRC)." http://image-net.org/challenges/LSVRC/.

"Intel's Efforts to Achieve a Responsible Minerals Supply Chain." Intel, May 2019. https://www.intel.com/content/www/us/en/corporate-responsibility/conflict-minerals-white-paper.html.

Irani, Lilly. "Difference and Dependence among Digital Workers: The Case of Amazon Mechanical Turk." *South Atlantic Quarterly* 114, no. 1 (2015): 225–34. https://doi.org/10.1215/00382876-2831665.

———. "The Hidden Faces of Automation." *XRDS* 23, no. 2 (2016): 34–37. https://doi.org/10.1145/3014390.

Izard, Carroll E. "The Many Meanings/Aspects of Emotion: Definitions,

Functions, Activation, and Regulation." *Emotion Review* 2, no. 4 (2010): 363–70. https://doi.org/10.1177/1754073910374661.

Jaton, Florian. "We Get the Algorithms of Our Ground Truths: Designing Referential Databases in Digital Image Processing." *Social Studies of Science* 47, no. 6 (2017): 811–40. https://doi.org/10.1177/0306312717730428.

Jin, Huafeng, and Shuo Wang. Voice-based determination of physical and emotional characteristics of users. US10096319B1, n.d.

Jobin, Anna, Marcello Ienca, and Effy Vayena. "The Global Landscape of AI Ethics Guidelines." *Nature Machine Intelligence* 1 (2019): 389–99. https://doi.org/10.1038/s42256-019-0088-2.

Jones, Nicola. "How to Stop Data Centres from Gobbling Up the World's Electricity." *Nature,* September 12, 2018. https://www.nature.com/articles/d41586-018-06610-y.

Joseph, George. "Data Company Directly Powers Immigration Raids in Workplace." *WNYC,* July 16, 2019. https://www.wnyc.org/story/palantir-directly-powers-ice-workplace-raids-emails-show/.

June, Laura. "YouTube Has a Fake Peppa Pig Problem." *The Outline,* March 16, 2017. https://theoutline.com/post/1239/youtube-has-a-fake-peppa-pig-problem.

Kafer, Alison. *Feminist, Queer, Crip.* Bloomington: Indiana University Press, 2013.

Kak, Amba, ed. "Regulating Biometrics: Global Approaches and Urgent Questions." AI Now Institute, September 1, 2020. https://ainowinstitute.org/regulatingbiometrics.html.

Kanade, Takeo. *Computer Recognition of Human Faces.* Basel: Birkhäuser Boston, 2013.

Kanade, T., J. F. Cohn, and Yingli Tian. "Comprehensive Database for Facial Expression Analysis." In *Proceedings Fourth IEEE International Conference on Automatic Face and Gesture Recognition,* 46–53. 2000. https://doi.org/10.1109/AFGR.2000.840611.

Kappas, A. "Smile When You Read This, Whether You Like It or Not: Conceptual Challenges to Affect Detection." *IEEE Transactions on Affective Computing* 1, no. 1 (2010): 38–41. https://doi.org/10.1109/T-AFFC.2010.6.

Katz, Lawrence F., and Alan B. Krueger. "The Rise and Nature of Alternative Work Arrangements in the United States, 1995–2015." *ILR Review* 72, no. 2 (2019): 382–416.

Keates, Nancy. "The Many Places Amazon CEO Jeff Bezos Calls Home." *Wall Street Journal,* January 9, 2019. https://www.wsj.com/articles/the-many-places-amazon-ceo-jeff-bezos-calls-home-1507204462.

Keel, Terence D. "Religion, Polygenism and the Early Science of Human Origins." *History of the Human Sciences* 26, no. 2 (2013): 3–32. https://doi.org/10.1177/0952695113482916.

Kelly, Kevin. *What Technology Wants.* New York: Penguin Books, 2011.

Kemeny, John, and Thomas Kurtz. "Dartmouth Timesharing." *Science* 162 (1968): 223–68.

Kendi, Ibram X. "A History of Race and Racism in America, in 24 Chapters." *New York Times,* February 22, 2017. https://www.nytimes.com/2017/02 /22/books/review/a-history-of-race-and-racism-in-america-in-24 -chapters.html.

Kerr, Dara. "Tech Workers Protest in SF to Keep Attention on Travel Ban." *CNET,* February 13, 2017. https://www.cnet.com/news/trump-immigra tion-ban-tech-workers-protest-no-ban-no-wall/.

Keyes, Os. "The Misgendering Machines: Trans/HCI Implications of Automatic Gender Recognition." In *Proceedings of the ACM on Human-Computer Interaction* 2, Issue CSCW (2018): art. 88. https://doi.org/10 .1145/3274357.

Kleinberg, Jon, et al. "Human Decisions and Machine Predictions." *Quarterly Journal of Economics* 133, no. 1 (2018): 237–93. https://doi.org/10.1093/qje /qjx032.

Klimt, Bryan, and Yiming Yang. "The Enron Corpus: A New Dataset for Email Classification Research." In *Machine Learning: ECML 2004,* edited by Jean-François Boulicat et al., 217–26. Berlin: Springer, 2004.

Klose, Alexander. *The Container Principle: How a Box Changes the Way We Think.* Translated by Charles Marcrum. Cambridge, Mass.: MIT Press, 2015.

Knight, Will. "Alpha Zero's 'Alien' Chess Shows the Power, and the Peculiarity, of AI." *MIT Technology Review,* December 8, 2017. https://www .technologyreview.com/s/609736/alpha-zeros-alien-chess-shows-the -power-and-the-peculiarity-of-ai/.

Kolbert, Elizabeth. "There's No Scientific Basis for Race—It's a Made-Up Label." *National Geographic,* March 12, 2018. https://www.national geographic.com/magazine/2018/04/race-genetics-science-africa/.

Krizhevsky, Alex, Ilya Sutskever, and Geoffrey E. Hinton. "ImageNet Classification with Deep Convolutional Neural Networks." *Communications of the ACM* 60, no. 6 (2017): 84–90. https://doi.org/10.1145/3065386.

Labban, Mazen. "Deterritorializing Extraction: Bioaccumulation and the Planetary Mine." *Annals of the Association of American Geographers* 104, no. 3 (2014): 560–76. https://www.jstor.org/stable/24537757.

Lakoff, George. *Women, Fire, and Dangerous Things: What Categories Reveal about the Mind.* Chicago: University of Chicago Press, 1987.

Lambert, Fred. "Breakdown of Raw Materials in Tesla's Batteries and Possible Breaknecks," electrek, November 1, 2016. https://electrek.co/2016/11/01 /breakdown-raw-materials-tesla-batteries-possible-bottleneck/.

Lapuschkin, Sebastian, et al. "Unmasking Clever Hans Predictors and Assessing What Machines Really Learn." *Nature Communications* 10, no. 1 (2019): 1–8. https://doi.org/10.1038/s41467-019-08987-4.

Latour, Bruno. "Tarde's Idea of Quantification." In *The Social after Gabriel Tarde: Debates and Assessments,* edited by Matei Candea, 147–64. New York: Routledge, 2010.

Lem, Stainslaw. "The First Sally (A), or Trurl's Electronic Bard." In *From Here to Forever,* vol. 4, *The Road to Science Fiction,* edited by James Gunn. Lanham, Md.: Scarecrow, 2003.

Leys, Ruth. *The Ascent of Affect: Genealogy and Critique.* Chicago: University of Chicago Press, 2017.

Li, Xiaochang. "Divination Engines: A Media History of Text Prediction." Ph.D. diss., New York University, 2017.

Libby, Sara. "Scathing Audit Bolsters Critics' Fears about Secretive State Gang Database." *Voice of San Diego,* August 11, 2016. https://www.voiceof sandiego.org/topics/public-safety/scathing-audit-bolsters-critics-fears -secretive-state-gang-database/.

Light, Jennifer S. "When Computers Were Women." *Technology and Culture* 40, no. 3 (1999): 455–83. https://www.jstor.org/stable/25147356.

Lingel, Jessa, and Kate Crawford. "Alexa, Tell Me about Your Mother: The History of the Secretary and the End of Secrecy." *Catalyst: Feminism, Theory, Technoscience* 6, no. 1 (2020). https://catalystjournal.org/index .php/catalyst/article/view/29949.

Liu, Zhiyi. "Chinese Mining Dump Could Hold Trillion-Dollar Rare Earth Deposit." China Dialogue, December 14, 2012. https://www.chinadialo gue.net/article/show/single/en/5495-Chinese-mining-dump-could-hold -trillion-dollar-rare-earth-deposit.

Lloyd, G. E. R. "The Development of Aristotle's Theory of the Classification of Animals." *Phronesis* 6, no. 1–2 (1961): 59–81. https://doi.org/10.1163/156 852861X00080.

Lo, Chris. "The False Monopoly: China and the Rare Earths Trade." *Mining Technology, Mining News and Views Updated Daily* (blog), August 19, 2015. https://www.mining-technology.com/features/featurethe-false -monopoly-china-and-the-rare-earths-trade-4646712/.

Locker, Melissa. "Microsoft, Duke, and Stanford Quietly Delete Databases with Millions of Faces." *Fast Company,* June 6, 2019. https://www.fast company.com/90360490/ms-celeb-microsoft-deletes-10m-faces-from -face-database.

Lorde, Audre. *The Master's Tools Will Never Dismantle the Master's House.* London: Penguin Classics, 2018.

Lucey, Patrick, et al. "The Extended Cohn-Kanade Dataset (CK+): A Complete Dataset for Action Unit and Emotion-Specified Expression." In *2010 IEEE Computer Society Conference on Computer Vision and Pat-*

tern Recognition — Workshops, 94–101. https://doi.org/10.1109/CVPRW
.2010.5543262.

Luxemburg, Rosa. "Practical Economies: Volume 2 of Marx's Capital." In *The Complete Works of Rosa Luxemburg,* edited by Peter Hudis, 421–60. London: Verso, 2013.

Lyons, M., et al. "Coding Facial Expressions with Gabor Wavelets." In *Proceedings Third IEEE International Conference on Automatic Face and Gesture Recognition,* 200–205. 1998. https://doi.org/10.1109/AFGR.1998.670949.

Lyotard, Jean François. "Presenting the Unpresentable: The Sublime." *Artforum,* April 1982.

Maass, Peter. "Summit Fever." *The Intercept* (blog), June 25, 2012. https:// www.documentcloud.org/documents/2088979-summit-fever.html.

Maass, Peter, and Beryl Lipton. "What We Learned." *MuckRock,* November 15, 2018. https://www.muckrock.com/news/archives/2018/nov/15/alpr -what-we-learned/.

MacKenzie, Donald A. *Inventing Accuracy: A Historical Sociology of Nuclear Missile Guidance.* Cambridge, Mass.: MIT Press, 2001.

"Magic from Invention." Brunel University London. https://www.brunel .ac.uk/research/Brunel-Innovations/Magic-from-invention.

Mahdawi, Arwa. "The Domino's 'Pizza Checker' Is Just the Beginning — Workplace Surveillance Is Coming for You." *Guardian,* October 15, 2019. https://www.theguardian.com/commentisfree/2019/oct/15/the-dominos -pizza-checker-is-just-the-beginning-workplace-surveillance-is-coming -for-you.

Marcus, Mitchell P., Mary Ann Marcinkiewicz, and Beatrice Santorini. "Building a Large Annotated Corpus of English: The Penn Treebank." *Computational Linguistics* 19, no. 2 (1993): 313–30. https://dl.acm.org/doi /abs/10.5555/972470.972475.

Markoff, John. "Pentagon Turns to Silicon Valley for Edge in Artificial Intelligence." *New York Times,* May 11, 2016. https://www.nytimes.com/2016/05 /12/technology/artificial-intelligence-as-the-pentagons-latest-weapon .html.

———. "Seeking a Better Way to Find Web Images." *New York Times,* November 19, 2012. https://www.nytimes.com/2012/11/20/science/for-web -images-creating-new-technology-to-seek-and-find.html.

———. "Skilled Work, without the Worker." *New York Times,* August 18, 2012. https://www.nytimes.com/2012/08/19/business/new-wave-of-ad ept-robots-is-changing-global-industry.html.

Martinage, Robert. "Toward a New Offset Strategy: Exploiting U.S. Long-Term Advantages to Restore U.S. Global Power Projection Capability." Washington, D.C.: Center for Strategic and Budgetary Assessments, 2014. https://csbaonline.org/uploads/documents/Offset-Strategy-Web.pdf.

Marx, Karl. *Das Kapital: A Critique of Political Economy.* Chicago: H. Regnery,
  1959.
———. *The Poverty of Philosophy.* New York: Progress, 1955.
Marx, Karl, and Friedrich Engels. *The Marx-Engels Reader,* edited by Rob-
  ert C. Tucker. 2nd ed. New York: W. W. Norton, 1978.
Marx, Paris. "Instead of Throwing Money at the Moon, Jeff Bezos Should
  Try Helping Earth." *NBC News,* May 15, 2019. https://www.nbcnews.com
  /think/opinion/jeff-bezos-blue-origin-space-colony-dreams-ignore
  -plight-millions-ncna1006026.
Masanet, Eric, Arman Shehabi, Nuoa Lei, Sarah Smith, and Jonathan Koomey.
  "Recalibrating Global Data Center Energy-Use Estimates." *Science* 367,
  no. 6481 (2020): 984–86.
Matney, Lucas. "More than 100 Million Alexa Devices Have Been Sold." *Tech-
  Crunch* (blog), January 4, 2019. http://social.techcrunch.com/2019/01/04
  /more-than-100-million-alexa-devices-have-been-sold/.
Mattern, Shannon. "Calculative Composition: The Ethics of Automating
  Design." In *The Oxford Handbook of Ethics of AI,* edited by Markus D.
  Dubber, Frank Pasquale, and Sunit Das, 572–92. Oxford: Oxford Univer-
  sity Press, 2020.
———. *Code and Clay, Data and Dirt: Five Thousand Years of Urban Media.*
  Minneapolis: University of Minnesota Press, 2017.
Maughan, Tim. "The Dystopian Lake Filled by the World's Tech Lust." BBC
  Future, April 2, 2015. https://www.bbc.com/future/article/20150402-the
  -worst-place-on-earth.
Mayhew, Claire, and Michael Quinlan. "Fordism in the Fast Food Indus-
  try: Pervasive Management Control and Occupational Health and Safety
  Risks for Young Temporary Workers." *Sociology of Health and Illness* 24,
  no. 3 (2002): 261–84. https://doi.org/10.1111/1467-9566.00294.
Mayr, Ernst. *The Growth of Biological Thought: Diversity, Evolution, and In-
  heritance.* Cambridge, Mass.: Harvard University Press, 1982.
Mbembé, Achille. *Critique of Black Reason.* Durham, N.C.: Duke University
  Press, 2017.
———. *Necropolitics.* Durham, N.C.: Duke University Press, 2019.
Mbembé, Achille, and Libby Meintjes. "Necropolitics." *Public Culture* 15,
  no. 1 (2003): 11–40. https://www.muse.jhu.edu/article/39984.
McCorduck, Pamela. *Machines Who Think: A Personal Inquiry into the His-
  tory and Prospects of Artificial Intelligence.* Natick, Mass.: A. K. Peters,
  2004.
McCurry, Justin. "Fukushima Disaster: Japanese Power Company Chiefs
  Cleared of Negligence." *Guardian,* September 19, 2019. https://www.the
  guardian.com/environment/2019/sep/19/fukushima-disaster-japanese
  -power-company-chiefs-cleared-of-negligence.

————. "Fukushima Nuclear Disaster: Former Tepco Executives Go on Trial." *Guardian,* June 30, 2017. https://www.theguardian.com/environ ment/2017/jun/30/fukushima-nuclear-crisis-tepco-criminal-trial-japan.

McDuff, Daniel, et al. "Affectiva-MIT Facial Expression Dataset (AM-FED): Naturalistic and Spontaneous Facial Expressions Collected 'In-the-Wild.'" In *2013 IEEE Conference on Computer Vision and Pattern Recognition Workshops,* 881–88. https://doi.org/10.1109/CVPRW.2013.130.

McIlwain, Charlton. *Black Software: The Internet and Racial Justice, from the AfroNet to Black Lives Matter.* New York: Oxford University Press, 2019.

McLuhan, Marshall. *Understanding Media: The Extensions of Man.* Reprint ed. Cambridge, Mass.: MIT Press, 1994.

McMillan, Graeme. "It's Not You, It's It: Voice Recognition Doesn't Recognize Women." *Time,* June 1, 2011. http://techland.time.com/2011/06/01 /its-not-you-its-it-voice-recognition-doesnt-recognize-women/.

McNamara, Robert S., and James G. Blight. *Wilson's Ghost: Reducing the Risk of Conflict, Killing, and Catastrophe in the 21st Century.* New York: Public Affairs, 2001.

McNeil, Joanne. "Two Eyes See More Than Nine." In *Jon Rafman: Nine Eyes,* edited by Kate Steinmann. Los Angeles: New Documents, 2016.

Mead, Margaret. Review of *Darwin and Facial Expression: A Century of Research in Review,* edited by Paul Ekman. *Journal of Communication* 25, no. 1 (1975): 209–40. https://doi.org/10.1111/j.1460-2466.1975.tb00574.x.

Meadows, Donella H., et al. *The Limits to Growth.* New York: Signet, 1972.

Menabrea, Luigi Federico, and Ada Lovelace. "Sketch of the Analytical Engine Invented by Charles Babbage." The Analytical Engine. https://www .fourmilab.ch/babbage/sketch.html.

Merler, Michele, et al. "Diversity in Faces." *ArXiv:1901.10436 [Cs],* April 8, 2019. http://arxiv.org/abs/1901.10436.

Metcalf, Jacob, and Kate Crawford. "Where Are Human Subjects in Big Data Research? The Emerging Ethics Divide." *Big Data and Society* 3, no. 1 (2016): 1–14. https://doi.org/10.1177/2053951716650211.

Metcalf, Jacob, Emanuel Moss, and danah boyd. "Owning Ethics: Corporate Logics, Silicon Valley, and the Institutionalization of Ethics." *International Quarterly* 82, no. 2 (2019): 449–76.

Meulen, Rob van der. "Gartner Says 8.4 Billion Connected 'Things' Will Be in Use in 2017, Up 31 Percent from 2016." *Gartner,* February 7, 2017. https://www.gartner.com/en/newsroom/press-releases/2017-02-07-gart ner-says-8-billion-connected-things-will-be-in-use-in-2017-up-31-per cent-from-2016.

Meyer, John W., and Ronald L. Jepperson. "The 'Actors' of Modern Society: The Cultural Construction of Social Agency." *Sociological Theory* 18, no. 1 (2000): 100–120. https://doi.org/10.1111/0735-2751.00090.

Mezzadra, Sandro, and Brett Neilson. "On the Multiple Frontiers of Extraction: Excavating Contemporary Capitalism." *Cultural Studies* 31, no. 2–3 (2017): 185–204. https://doi.org/10.1080/09502386.2017.1303425.

Michalski, Ryszard S. "Pattern Recognition as Rule-Guided Inductive Inference." *IEEE Transactions on Pattern Analysis Machine Intelligence* 2, no. 4 (1980): 349–61. https://doi.org/10.1109/TPAMI.1980.4767034.

Michel, Arthur Holland. *Eyes in the Sky: The Secret Rise of Gorgon Stare and How It Will Watch Us All.* Boston: Houghton Mifflin Harcourt, 2019.

Mikel, Betsy. "WeWork Just Made a Disturbing Acquisition; It Raises a Lot of Flags about Workers' Privacy." Inc.com, February 17, 2019. https://www.inc.com/betsy-mikel/wework-is-trying-a-creepy-new-strategy-it-just-might-signal-end-of-workplace-as-we-know-it.html.

Mirzoeff, Nicholas. *The Right to Look: A Counterhistory of Visuality.* Durham, N.C.: Duke University Press, 2011.

Mitchell, Margaret, et al. "Model Cards for Model Reporting." In *FAT* '19: Proceedings of the Conference on Fairness, Accountability, and Transparency,* 220–29. Atlanta: ACM Press, 2019. https://doi.org/10.1145/3287560.3287596.

Mitchell, Paul Wolff. "The Fault in His Seeds: Lost Notes to the Case of Bias in Samuel George Morton's Cranial Race Science." *PLOS Biology* 16, no. 10 (2018): e2007008. https://doi.org/10.1371/journal.pbio.2007008.

Mitchell, Tom M. "The Need for Biases in Learning Generalizations." Working paper, Rutgers University, May 1980.

Mitchell, W. J. T. *Picture Theory: Essays on Verbal and Visual Representation.* Chicago.: University of Chicago Press, 1994.

Mittelstadt, Brent. "Principles Alone Cannot Guarantee Ethical AI." *Nature Machine Intelligence* 1, no. 11 (2019): 501–7. https://doi.org/10.1038/s42256-019-0114-4.

Mohamed, Shakir, Marie-Therese Png, and William Isaac. "Decolonial AI: Decolonial Theory as Sociotechnical Foresight in Artificial Intelligence." *Philosophy and Technology* (2020): 405. https://doi.org/10.1007/s13347-020-00405-8.

Moll, Joana. "CO2GLE." http://www.janavirgin.com/CO2/.

Molnar, Phillip, Gary Robbins, and David Pierson. "Cutting Edge: Apple's Purchase of Emotient Fuels Artificial Intelligence Boom in Silicon Valley." *Los Angeles Times,* January 17, 2016. https://www.latimes.com/business/technology/la-fi-cutting-edge-facial-recognition-20160117-story.html.

Morris, David Z. "Major Advertisers Flee YouTube over Videos Exploiting Children." *Fortune,* November 26, 2017. https://fortune.com/2017/11/26/advertisers-flee-youtube-child-exploitation/.

Morton, Timothy. *Hyperobjects: Philosophy and Ecology after the End of the World.* Minneapolis: University of Minnesota Press, 2013.

Mosco, Vincent. *To the Cloud: Big Data in a Turbulent World*. Boulder, Colo.: Paradigm, 2014.

Müller-Maguhn, Andy, et al. "The NSA Breach of Telekom and Other German Firms." *Spiegel,* September 14, 2014. https://www.spiegel.de/international/world/snowden-documents-indicate-nsa-has-breached-deutsche-telekom-a-991503.html.

Mumford, Lewis. "The First Megamachine." *Diogenes* 14, no. 55 (1966): 1–15. https://doi.org/10.1177/039219216601405501.

———. *The Myth of the Machine*. Vol. 1: *Technics and Human Development*. New York: Harcourt Brace Jovanovich, 1967.

———. *Technics and Civilization*. Chicago: University of Chicago Press, 2010.

Murgia, Madhumita, and Max Harlow. "Who's Using Your Face? The Ugly Truth about Facial Recognition." *Financial Times,* April 19, 2019. https://www.ft.com/content/cf19b956-60a2-11e9-b285-3acd5d43599e.

Muse, Abdi. "Organizing Tech." AI Now 2019 Symposium, AI Now Institute, 2019. https://ainowinstitute.org/symposia/2019-symposium.html.

Nakashima, Ellen, and Joby Warrick. "For NSA Chief, Terrorist Threat Drives Passion to 'Collect It All.'" *Washington Post,* July 14, 2013. https://www.washingtonpost.com/world/national-security/for-nsa-chief-terrorist-threat-drives-passion-to-collect-it-all/2013/07/14/3d26ef80-ea49-11e2-a301-ea5a8116d211_story.html.

NASA. "Outer Space Treaty of 1967." NASA History, 1967. https://history.nasa.gov/1967treaty.html.

Nassar, Nedal, et al. "Evaluating the Mineral Commodity Supply Risk of the US Manufacturing Sector." *Science Advances* 6, no. 8 (2020): eaa8647. https://www.doi.org/10.1126/sciadv.aay8647.

Natarajan, Prem. "Amazon and NSF Collaborate to Accelerate Fairness in AI Research." *Alexa Blogs* (blog), March 25, 2019. https://developer.amazon.com/blogs/alexa/post/1786ea03-2e55-4a93-9029-5df88c200ac1/amazon-and-nsf-collaborate-to-accelerate-fairness-in-ai-research.

National Institute of Standards and Technology (NIST). "Special Database 32—Multiple Encounter Dataset (MEDS)." https://www.nist.gov/itl/iad/image-group/special-database-32-multiple-encounter-dataset-meds.

Nedlund, Evelina. "Apple Card Is Accused of Gender Bias; Here's How That Can Happen." *CNN,* November 12, 2019. https://edition.cnn.com/2019/11/12/business/apple-card-gender-bias/index.html.

Negroni, Christine. "How to Determine the Power Rating of Your Gadget's Batteries." *New York Times,* December 26, 2016. https://www.nytimes.com/2016/12/26/business/lithium-ion-battery-airline-safety.html.

"Neighbors by Ring: Appstore for Android." Amazon. https://www.amazon.com/Ring-Neighbors-by/dp/B07V7K49QT.

Nelson, Alondra. *The Social Life of DNA: Race, Reparations, and Reconciliation after the Genome.* Boston: Beacon, 2016.

Nelson, Alondra, Thuy Linh N. Tu, and Alicia Headlam Hines. "Introduction: Hidden Circuits." In *Technicolor: Race, Technology, and Everyday Life,* edited by Alondra Nelson, Thuy Linh N. Tu, and Alicia Headlam Hines, 1–12. New York: New York University Press 2001.

Nelson, Francis W., and Henry Kucera. *Brown Corpus Manual: Manual of Information to Accompany a Standard Corpus of Present-Day Edited American English for Use with Digital Computers.* Providence, R.I.: Brown University, 1979. http://icame.uib.no/brown/bcm.html.

Nelson, Robin. "Racism in Science: The Taint That Lingers." *Nature* 570 (2019): 440–41. https://doi.org/10.1038/d41586-019-01968-z.

Newman, Lily Hay. "Internal Docs Show How ICE Gets Surveillance Help From Local Cops." *Wired,* March 13, 2019. https://www.wired.com/story/ice-license-plate-surveillance-vigilant-solutions/.

Nielsen, Kim E. *A Disability History of the United States.* Boston: Beacon, 2012.

Nietzsche, Friedrich. *Sämtliche Werke.* Vol. 11. Berlin: de Gruyter, 1980.

Nilsson, Nils J. *The Quest for Artificial Intelligence: A History of Ideas and Achievements.* New York: Cambridge University Press, 2009.

Nilsson, Patricia. "How AI Helps Recruiters Track Jobseekers' Emotions." *Financial Times,* February 28, 2018. https://www.ft.com/content/e2e85644-05be-11e8-9650-9c0ad2d7c5b5.

Noble, Safiya Umoja. *Algorithms of Oppression: How Search Engines Reinforce Racism.* New York: NYU Press, 2018.

"NSA Phishing Tactics and Man in the Middle Attacks." *The Intercept* (blog), March 12, 2014. https://theintercept.com/document/2014/03/12/nsa-phishing-tactics-man-middle-attacks/.

"Off Now: How Your State Can Help Support the Fourth Amendment." OffNow.org. *https://s3.amazonaws.com/TAChandbooks/OffNow-Handbook.pdf.*

Ohm, Paul. "Don't Build a Database of Ruin." *Harvard Business Review,* August 23, 2012. https://hbr.org/2012/08/dont-build-a-database-of-ruin.

Ohtake, Miyoko. "Psychologist Paul Ekman Delights at Exploratorium." *WIRED,* January 28, 2008. https://www.wired.com/2008/01/psychologist-pa/.

O'Neil, Cathy. *Weapons of Math Destruction: How Big Data Increases Inequality and Threatens Democracy.* New York: Crown, 2016.

O'Neill, Gerard K. *The High Frontier: Human Colonies in Space.* 3rd ed. Burlington, Ont.: Apogee Books, 2000.

"One-Year Limited Warranty for Amazon Devices or Accessories." Amazon.

https://www.amazon.com/gp/help/customer/display.html?nodeId=20
1014520.

"An Open Letter." https://ethuin.files.wordpress.com/2014/09/09092014
-open-letter-final-and-list.pdf.

Organizing Tech. Video, AI Now Institute, 2019. https://www.youtube.com
/watch?v=jLeOyIS1jwc&feature=emb_title.

Osumi, Magdalena. "Former Tepco Executives Found Not Guilty of Criminal Negligence in Fukushima Nuclear Disaster." *Japan Times Online,* September 19, 2019. https://www.japantimes.co.jp/news/2019/09/19/natio
nal/crime-legal/tepco-trio-face-tokyo-court-ruling-criminal-case-
stemming-fukushima-nuclear-disaster/.

"Our Mission." Blue Origin. https://www-dev.blueorigin.com/our-mission.

Paglen, Trevor. "Operational Images." *e-flux,* November 2014. https://www
.e-flux.com/journal/59/61130/operational-images/.

Palantir. "Palantir Gotham." https://palantir.com/palantir-gotham/index
.html.

"Palantir and Cambridge Analytica: What Do We Know?" *WikiTribune,*
March 27, 2018. https://www.wikitribune.com/wt/news/article/58386/.

Pande, Vijay. "Artificial Intelligence's 'Black Box' Is Nothing to Fear." *New
York Times,* January 25, 2018. https://www.nytimes.com/2018/01/25/opin
ion/artificial-intelligence-black-box.html.

Papert, Seymour A. "The Summer Vision Project." July 1, 1966. https://dspace
.mit.edu/handle/1721.1/6125.

Parikka, Jussi. *A Geology of Media.* Minneapolis: University of Minnesota
Press, 2015.

Pasquale, Frank. *The Black Box Society: The Secret Algorithms That Control
Money and Information.* Cambridge, Mass.: Harvard University Press,
2015.

Patterson, Scott, and Alexandra Wexler. "Despite Cleanup Vows, Smartphones and Electric Cars Still Keep Miners Digging by Hand in Congo."
Wall Street Journal, September 13, 2018. https://www.wsj.com/articles
/smartphones-electric-cars-keep-miners-digging-by-hand-in-congo
-1536835334.

Paul Ekman Group. https://www.paulekman.com/.

Pellerin, Cheryl. "Deputy Secretary: Third Offset Strategy Bolsters America's
Military Deterrence." Washington, D.C.: U.S. Department of Defense,
October 31, 2016. https://www.defense.gov/Explore/News/Article/Arti
cle/991434/deputy-secretary-third-offset-strategy-bolsters-americas
-military-deterrence/.

Perez, Sarah. "Microsoft Silences Its New A.I. Bot Tay, after Twitter Users
Teach It Racism [Updated]." *TechCrunch* (blog), March 24, 2016. http://

social.techcrunch.com/2016/03/24/microsoft-silences-its-new-a-i-bot
-tay-after-twitter-users-teach-it-racism/.

Pfungst, Oskar. *Clever Hans (The Horse of Mr. von Osten): A Contribution to Experimental Animal and Human Psychology.* Translated by Carl L. Rahn. New York: Henry Holt, 1911.

Phillips, P. Jonathon, Patrick J. Rauss, and Sandor Z. Der. "FERET (Face Recognition Technology) Recognition Algorithm Development and Test Results." Adelphi, Md.: Army Research Laboratory, October 1996. https://apps.dtic.mil/dtic/tr/fulltext/u2/a315841.pdf.

Picard, Rosalind. "Affective Computing Group." MIT Media Lab. https://affect.media.mit.edu/.

Pichai, Sundar. "AI at Google: Our Principles." Google, June 7, 2018. https://blog.google/technology/ai/ai-principles/.

Plumwood, Val. "The Politics of Reason: Towards a Feminist Logic." *Australasian Journal of Philosophy* 71, no. 4 (1993): 436–62. https://doi.org/10.1080/00048409312345432.

Poggio, Tomaso, et al. "Why and When Can Deep—but not Shallow—Networks Avoid the Curse of Dimensionality: A Review." *International Journal of Automation and Computing* 14, no. 5 (2017): 503–19. https://link.springer.com/article/10.1007/s11633-017-1054-2.

Pontin, Jason. "Artificial Intelligence, with Help from the Humans." *New York Times,* March 25, 2007. https://www.nytimes.com/2007/03/25/business/yourmoney/25Stream.html.

Pontin, Mark Williams. "Lie Detection." *MIT Technology Review,* April 21, 2009. https://www.technologyreview.com/s/413133/lie-detection/.

Powell, Corey S. "Jeff Bezos Foresees a Trillion People Living in Millions of Space Colonies." *NBC News,* May 15, 2019. https://www.nbcnews.com/mach/science/jeff-bezos-foresees-trillion-people-living-millions-space-colonies-here-ncna1006036.

"Powering the Cloud: How China's Internet Industry Can Shift to Renewable Energy." Greenpeace, September 9, 2019. https://storage.googleapis.com/planet4-eastasia-stateless/2019/11/7bfe9069-7bfe9069-powering-the-cloud-_-english-briefing.pdf.

Pratt, Mary Louise. "Arts of the Contact Zone." *Profession, Ofession* (1991): 33–40.

———. *Imperial Eyes: Travel Writing and Transculturation.* 2nd ed. London: Routledge, 2008.

Priest, Dana. "NSA Growth Fueled by Need to Target Terrorists." *Washington Post,* July 21, 2013. https://www.washingtonpost.com/world/national-security/nsa-growth-fueled-by-need-to-target-terrorists/2013/07/21/24c93cf4-f0b1-11e2-bed3-b9b6fe264871_story.html.

Pryzbylski, David J. "Changes Coming to NLRB's Stance on Company

E-Mail Policies?" *National Law Review,* August 2, 2018. https://www.nat lawreview.com/article/changes-coming-to-nlrb-s-stance-company-e -mail-policies.

Puar, Jasbir K. *Terrorist Assemblages: Homonationalism in Queer Times.* 2nd ed. Durham, N.C.: Duke University Press, 2017.

Pugliese, Joseph. "Death by Metadata: The Bioinformationalisation of Life and the Transliteration of Algorithms to Flesh." In *Security, Race, Biopower: Essays on Technology and Corporeality,* edited by Holly Randell-Moon and Ryan Tippet, 3–20. London: Palgrave Macmillan, 2016.

Puschmann, Cornelius, and Jean Burgess. "Big Data, Big Questions: Metaphors of Big Data." *International Journal of Communication* 8 (2014): 1690–1709.

Qiu, Jack. *Goodbye iSlave: A Manifesto for Digital Abolition.* Urbana: University of Illinois Press, 2016.

Qiu, Jack, Melissa Gregg, and Kate Crawford. "Circuits of Labour: A Labour Theory of the iPhone Era." *TripleC: Communication, Capitalism and Critique* 12, no. 2 (2014). https://doi.org/10.31269/triplec.v12i2.540.

"Race after Technology, Ruha Benjamin." Meeting minutes, Old Guard of Princeton, N.J., November 14, 2018. https://www.theoldguardofprince ton.org/11-14-2018.html.

Raji, Inioluwa Deborah, and Joy Buolamwini. "Actionable Auditing: Investigating the Impact of Publicly Naming Biased Performance Results of Commercial AI Products." In *Proceedings of the 2019 AAAI/ACM Conference on AI, Ethics, and Society,* 429–35. 2019.

Raji, Inioluwa Deborah, Timnit Gebru, Margaret Mitchell, Joy Buolamwini, Joonseok Lee, and Emily Denton. "Saving Face: Investigating the Ethical Concerns of Facial Recognition Auditing." In *Proceedings of the AAAI/ ACM Conference on AI, Ethics, and Society,* 145–51. 2020.

Ramachandran, Vilayanur S., and Diane Rogers-Ramachandran. "Aristotle's Error." *Scientific American,* March 1, 2010. https://doi.org/10.1038/scien tificamericanmind0310-20.

Rankin, Joy Lisi. *A People's History of Computing in the United States.* Cambridge, Mass.: Harvard University Press, 2018.

———. "Remembering the Women of the Mathematical Tables Project." *The New Inquiry* (blog), March 14, 2019. https://thenewinquiry.com /blog/remembering-the-women-of-the-mathematical-tables-project/.

Rehmann, Jan. "Taylorism and Fordism in the Stockyards." In *Max Weber: Modernisation as Passive Revolution,* 24–29. Leiden, Netherlands: Brill, 2015.

Reichhardt, Tony. "First Photo from Space." *Air and Space Magazine,* October 24, 2006. https://www.airspacemag.com/space/the-first-photo-from -space-13721411/.

Rein, Hanno, Daniel Tamayo, and David Vokrouhlicky. "The Random Walk of Cars and Their Collision Probabilities with Planets." *Aerospace* 5, no. 2 (2018): 57. https://doi.org/10.3390/aerospace5020057.

"Responsible Minerals Policy and Due Diligence." Philips. https://www .philips.com/a-w/about/company/suppliers/supplier-sustainability/our -programs/responsible-sourcing-of-minerals.html.

"Responsible Minerals Sourcing." Dell. https://www.dell.com/learn/us/en /uscorp1/conflict-minerals?s=corp.

Revell, Timothy. "Google DeepMind's NHS Data Deal 'Failed to Comply' with Law." *New Scientist,* July 3, 2017. https://www.newscientist.com /article/2139395-google-deepminds-nhs-data-deal-failed-to-comply -with-law/.

Rhue, Lauren. "Racial Influence on Automated Perceptions of Emotions." November 9, 2018. https://dx.doi.org/10.2139/ssrn.3281765.

Richardson, Rashida, Jason M. Schultz, and Kate Crawford. "Dirty Data, Bad Predictions: How Civil Rights Violations Impact Police Data, Predictive Policing Systems, and Justice." *NYU Law Review Online* 94, no. 15 (2019): 15–55. https://www.nyulawreview.org/wp-content/uploads/2019/04/NY ULawReview-94-Richardson-Schultz-Crawford.pdf.

Richardson, Rashida, Jason M. Schultz, and Vincent M. Southerland. "Litigating Algorithms: 2019 US Report." AI Now Institute, September 2019. https://ainowinstitute.org/litigatingalgorithms-2019-us.pdf.

Risen, James, and Laura Poitras. "N.S.A. Report Outlined Goals for More Power." *New York Times,* November 22, 2013. https://www.nytimes.com /2013/11/23/us/politics/nsa-report-outlined-goals-for-more-power.html.

Robbins, Martin. "How Can Our Future Mars Colonies Be Free of Sexism and Racism?" *Guardian,* May 6, 2015. https://www.theguardian.com/sci ence/the-lay-scientist/2015/may/06/how-can-our-future-mars-colonies -be-free-of-sexism-and-racism.

Roberts, Dorothy. *Fatal Invention: How Science, Politics, and Big Business Re-Create Race in the Twenty-First Century.* New York: New Press, 2011.

Roberts, Sarah T. *Behind the Screen: Content Moderation in the Shadows of Social Media.* New Haven: Yale University Press, 2019.

Romano, Benjamin. "Suits Allege Amazon's Alexa Violates Laws by Recording Children's Voices without Consent." *Seattle Times,* June 12, 2019. https://www.seattletimes.com/business/amazon/suit-alleges-amazons -alexa-violates-laws-by-recording-childrens-voices-without-consent/.

Romm, Tony. "U.S. Government Begins Asking Foreign Travelers about Social Media." *Politico,* December 22, 2016. https://www.politico.com/story /2016/12/foreign-travelers-social-media-232930.

Rouast, Philipp V., Marc Adam, and Raymond Chiong. "Deep Learning for Human Affect Recognition: Insights and New Developments." In *IEEE*

Transactions on Affective Computing, 2019, 1. https://doi.org/10.1109/TAF
  FC.2018.2890471.

"Royal Free–Google DeepMind Trial Failed to Comply with Data Protection
  Law." Information Commissioner's Office, July 3, 2017. https://ico.org.uk
  /about-the-ico/news-and-events/news-and-blogs/2017/07/royal-free
  -google-deepmind-trial-failed-to-comply-with-data-protection-law/.

Russell, Andrew. *Open Standards and the Digital Age: History, Ideology, and
  Networks.* New York: Cambridge University Press, 2014.

Russell, James A. "Is There Universal Recognition of Emotion from Facial
  Expression? A Review of the Cross-Cultural Studies." *Psychological Bulletin* 115, no. 1 (1994): 102–41. https://doi.org/10.1037/0033-2909.115.1.102.

Russell, Stuart J., and Peter Norvig. *Artificial Intelligence: A Modern Approach.* 3rd ed. Upper Saddle River, N.J.: Pearson, 2010.

Sadowski, Jathan. "When Data Is Capital: Datafication, Accumulation, and
  Extraction." *Big Data and Society* 6, no. 1 (2019): 1–12. https://doi.org/10
  .1177/2053951718820549.

Sadowski, Jathan. "Potemkin AI." *Real Life,* August 6, 2018.

Sample, Ian. "What Is the Internet? 13 Key Questions Answered." *Guardian,*
  October 22, 2018. https://www.theguardian.com/technology/2018/oct/22
  /what-is-the-internet-13-key-questions-answered.

Sánchez-Monedero, Javier, and Lina Dencik. "The Datafication of the Workplace." Working paper, Data Justice Lab, Cardiff University, May 9, 2019.
  https://datajusticeproject.net/wp-content/uploads/sites/30/2019/05
  /Report-The-datafication-of-the-workplace.pdf.

Sanville, Samantha. "Towards Humble Geographies." *Area* (2019): 1–9. https://
  doi.org/10.1111/area.12664.

Satisky, Jake. "A Duke Study Recorded Thousands of Students' Faces; Now
  They're Being Used All over the World." *Chronicle,* June 12, 2019. https://
  www.dukechronicle.com/article/2019/06/duke-university-facial-recogni
  tion-data-set-study-surveillance-video-students-china-uyghur.

Scahill, Jeremy, and Glenn Greenwald. "The NSA's Secret Role in the U.S.
  Assassination Program." *The Intercept* (blog), February 10, 2014. https://
  theintercept.com/2014/02/10/the-nsas-secret-role/.

Schaake, Marietje. "What Principles Not to Disrupt: On AI and Regulation." *Medium* (blog), November 5, 2019. https://medium.com/@marietje
  .schaake/what-principles-not-to-disrupt-on-ai-and-regulation-cabbd92
  fd30e.

Schaffer, Simon. "Babbage's Calculating Engines and the Factory System."
  Réseaux: Communication — Technologie — Société 4, no. 2 (1996): 271–98.
  https://doi.org/10.3406/reso.1996.3315.

Scharmen, Fred. *Space Settlements.* New York: Columbia University Press,
  2019.

Scharre, Paul, et al. "Eric Schmidt Keynote Address at the Center for a New American Security Artificial Intelligence and Global Security Summit." Center for a New American Security, November 13, 2017. https://www .cnas.org/publications/transcript/eric-schmidt-keynote-address-at-the -center-for-a-new-american-security-artificial-intelligence-and-global -security-summit.

Scheuerman, Morgan Klaus, et al. "How We've Taught Algorithms to See Identity: Constructing Race and Gender in Image Databases for Facial Analysis." *Proceedings of the ACM on Human-Computer Interaction* 4, issue CSCW1 (2020): 1–35. https://doi.org/10.1145/3392866.

Scheyder, Ernest. "Tesla Expects Global Shortage of Electric Vehicle Battery Minerals." *Reuters*, May 2, 2019. https://www.reuters.com/article/us-usa -lithium-electric-tesla-exclusive-idUSKCN1S81QS.

Schlanger, Zoë. "If Shipping Were a Country, It Would Be the Sixth-Biggest Greenhouse Gas Emitter." *Quartz*, April 17, 2018. https://qz.com/1253874 /if-shipping-were-a-country-it-would-the-worlds-sixth-biggest-green house-gas-emitter/.

Schmidt, Eric. "I Used to Run Google; Silicon Valley Could Lose to China." *New York Times*, February 27, 2020. https://www.nytimes.com/2020/02 /27/opinion/eric-schmidt-ai-china.html.

Schneier, Bruce. "Attacking Tor: How the NSA Targets Users' Online Anonymity." *Guardian*, October 4, 2013. https://www.theguardian.com/world /2013/oct/04/tor-attacks-nsa-users-online-anonymity.

Schwartz, Oscar. "Don't Look Now: Why You Should Be Worried about Machines Reading Your Emotions." *Guardian*, March 6, 2019. https://www .theguardian.com/technology/2019/mar/06/facial-recognition-software -emotional-science.

Scott, James C. *Seeing Like a State: How Certain Schemes to Improve the Human Condition Have Failed.* New Haven: Yale University Press, 1998.

Sedgwick, Eve Kosofsky. *Touching Feeling: Affect, Pedagogy, Performativity.* Durham, N.C.: Duke University Press, 2003.

Sedgwick, Eve Kosofsky, Adam Frank, and Irving E. Alexander, eds. *Shame and Its Sisters: A Silvan Tomkins Reader.* Durham, N.C.: Duke University Press, 1995.

Sekula, Allan. "The Body and the Archive." *October* 39 (1986): 3–64. https:// doi.org/10.2307/778312.

Senechal, Thibaud, Daniel McDuff, and Rana el Kaliouby. "Facial Action Unit Detection Using Active Learning and an Efficient Non-Linear Kernel Approximation." In *2015 IEEE International Conference on Computer Vision Workshop (ICCVW)*, 10–18. https://doi.org/10.1109/ICCVW.2015.11.

Senior, Ana. "John Hancock Leaves Traditional Life Insurance Model Behind to Incentivize Longer, Healthier Lives." Press release, John Hancock, September 19, 2018.

Seo, Sungyong, et al. "Partially Generative Neural Networks for Gang Crime Classification with Partial Information." In *Proceedings of the 2018 AAAI/ ACM Conference on AI, Ethics, and Society*, 257–263. https://doi.org/10.1145 /3278721.3278758.

Shaer, Matthew. "The Asteroid Miner's Guide to the Galaxy." *Foreign Policy* (blog), April 28, 2016. https://foreignpolicy.com/2016/04/28/the-asteroid -miners-guide-to-the-galaxy-space-race-mining-asteroids-planetary-re search-deep-space-industries/.

Shane, Scott, and Daisuke Wakabayashi. "'The Business of War': Google Employees Protest Work for the Pentagon." *New York Times*, April 4, 2018. https://www.nytimes.com/2018/04/04/technology/google-letter-ceo -pentagon-project.html.

Shankleman, Jessica, et al. "We're Going to Need More Lithium." *Bloomberg*, September 7, 2017. https://www.bloomberg.com/graphics/2017-lithium- battery-future/.

SHARE Foundation. "Serbian Government Is Implementing Unlawful Video Surveillance with Face Recognition in Belgrade." Policy brief, undated. https://www.sharefoundation.info/wp-content/uploads/Serbia-Video -Surveillance-Policy-brief-final.pdf.

Siebers, Tobin. *Disability Theory*. Ann Arbor: University of Michigan Press, 2008.

Siegel, Erika H., et al. "Emotion Fingerprints or Emotion Populations? A Meta-Analytic Investigation of Autonomic Features of Emotion Categories." *Psychological Bulletin* 144, no. 4 (2018): 343–93. https://doi.org /10.1037/bul0000128.

Silberman, M. S., et al. "Responsible Research with Crowds: Pay Crowd-workers at Least Minimum Wage." *Communications of the ACM* 61, no. 3 (2018): 39–41. https://doi.org/10.1145/3180492.

Silver, David, et al. "Mastering the Game of Go without Human Knowledge." *Nature* 550 (2017): 354–59. https://doi.org/10.1038/nature24270.

Simmons, Brandon. "Rekor Software Adds License Plate Reader Technology to Home Surveillance, Causing Privacy Concerns." *WKYC*, January 31, 2020. https://www.wkyc.com/article/tech/rekor-software-adds-license -plate-reader-technology-to-home-surveillance-causing-privacy-con cerns/95-7c9834d9-5d54-4081-b983-b2e6142a3213.

Simpson, Cam. "The Deadly Tin inside Your Smartphone." *Bloomberg*, August 24, 2012. https://www.bloomberg.com/news/articles/2012-08-23/the -deadly-tin-inside-your-smartphone.

Singh, Amarjot. *Eye in the Sky: Real-Time Drone Surveillance System (DSS)*

for Violent Individuals Identification. Video, June 2, 2018. https://www
.youtube.com/watch?time_continue=1&v=zYypJPJipYc.

"SKYNET: Courier Detection via Machine Learning." *The Intercept* (blog),
May 8, 2015. https://theintercept.com/document/2015/05/08/skynet
-courier/.

Sloane, Garett. "Online Ads for High-Paying Jobs Are Targeting Men More
Than Women." *AdWeek* (blog), July 7, 2015. https://www.adweek.com
/digital/seemingly-sexist-ad-targeting-offers-more-men-women-high
-paying-executive-jobs-165782/.

Smith, Adam. *An Inquiry into the Nature and Causes of the Wealth of Nations.*
Chicago: University of Chicago Press, 1976.

Smith, Brad. "Microsoft Will Be Carbon Negative by 2030." *Official Microsoft
Blog* (blog), January 20, 2020. https://blogs.microsoft.com/blog/2020/01
/16/microsoft-will-be-carbon-negative-by-2030/.

———. "Technology and the US Military." *Microsoft on the Issues* (blog),
October 26, 2018. https://blogs.microsoft.com/on-the-issues/2018/10/26
/technology-and-the-us-military/.

"Snowden Archive: The SIDtoday Files." *The Intercept* (blog), May 29, 2019.
https://theintercept.com/snowden-sidtoday/.

Solon, Olivia. "Facial Recognition's 'Dirty Little Secret': Millions of Online
Photos Scraped without Consent." NBC News, March 12, 2019. https://
www.nbcnews.com/tech/internet/facial-recognition-s-dirty-little-secret
-millions-online-photos-scraped-n981921.

Souriau, Étienne. *The Different Modes of Existence.* Translated by Erik Bera-
nek and Tim Howles. Minneapolis: University of Minnesota Press, 2015.

Spangler, Todd. "Listen to the Big Ticket with Marc Malkin." IHeartRadio,
May 3, 2019. https://www.iheart.com/podcast/28955447/.

Spargo, John. *Syndicalism, Industrial Unionism, and Socialism* [1913]. St.
Petersburg, Fla.: Red and Black, 2009.

Specht, Joshua. *Red Meat Republic: A Hoof-to-Table History of How Beef
Changed America.* Princeton, N.J.: Princeton University Press, 2019.

Standage, Tom. *The Turk: The Life and Times of the Famous Eighteenth-
Century Chess-Playing Machine.* New York: Walker, 2002.

Stark, Luke. "Facial Recognition Is the Plutonium of AI." *XRDS: Crossroads,
The ACM Magazine for Students* 25, no. 3 (2019). https://doi.org/10.1145
/3313129.

Stark, Luke, and Anna Lauren Hoffmann. "Data Is the New What? Popular
Metaphors and Professional Ethics in Emerging Data Culture." *Journal of
Cultural Analytics* 1, no. 1 (2019). https://doi.org/10.22148/16.036.

Starosielski, Nicole. *The Undersea Network.* Durham, N.C.: Duke University
Press, 2015.

Steadman, Philip. "Samuel Bentham's Panopticon." *Journal of Bentham Studies* 2 (2012): 1–30. https://doi.org/10.14324/111.2045-757X.044.

Steinberger, Michael. "Does Palantir See Too Much?" *New York Times Magazine*, October 21, 2020. https://www.nytimes.com/interactive/2020/10/21/magazine/palantir-alex-karp.html.

Stewart, Ashley, and Nicholas Carlson. "The President of Microsoft Says It Took Its Bid for the $10 Billion JEDI Cloud Deal as an Opportunity to Improve Its Tech—and That's Why It Beat Amazon." *Business Insider,* January 23, 2020. https://www.businessinsider.com/brad-smith-microsofts-jedi-win-over-amazon-was-no-surprise-2020-1.

Stewart, Russell. *Brainwash Dataset.* Stanford Digital Repository, 2015. https://purl.stanford.edu/sx925dc9385.

Stoller, Bill. "Why the Northern Virginia Data Center Market Is Bigger Than Most Realize." Data Center Knowledge, February 14, 2019. https://www.datacenterknowledge.com/amazon/why-northern-virginia-data-center-market-bigger-most-realize.

Strand, Ginger Gail. "Keyword: Evil." *Harper's Magazine,* March 2008. https://harpers.org/archive/2008/03/keyword/.

"A Strategy for Surveillance Powers." *New York Times,* February 23, 2012. https://www.nytimes.com/interactive/2013/11/23/us/politics/23nsa-sigint-strategy-document.html.

"Street Homelessness." San Francisco Department of Homelessness and Supportive Housing. http://hsh.sfgov.org/street-homelessness/.

Strubell, Emma, Ananya Ganesh, and Andrew McCallum. "Energy and Policy Considerations for Deep Learning in NLP." *ArXiv:1906.02243 [Cs],* June 5, 2019. http://arxiv.org/abs/1906.02243.

Suchman, Lucy. "Algorithmic Warfare and the Reinvention of Accuracy." *Critical Studies on Security* (2020): n. 18. https://doi.org/10.1080/21624887.2020.1760587.

Sullivan, Mark. "Fact: Apple Reveals It Has 900 Million iPhones in the Wild." *Fast Company,* January 29, 2019. https://www.fastcompany.com/90298944/fact-apple-reveals-it-has-900-million-iphones-in-the-wild.

Sutton, Rich. "The Bitter Lesson." March 13, 2019. http://www.incomplete-ideas.net/IncIdeas/BitterLesson.html.

Swinhoe, Dan. "What Is Spear Phishing? Why Targeted Email Attacks Are So Difficult to Stop." CSO Online, January 21, 2019. https://www.csoonline.com/article/3334617/what-is-spear-phishing-why-targeted-email-attacks-are-so-difficult-to-stop.html.

Szalai, Jennifer. "How the 'Temp' Economy Became the New Normal." *New York Times,* August 22, 2018. https://www.nytimes.com/2018/08/22/books/review-temp-louis-hyman.html.

Tani, Maxwell. "The Intercept Shuts Down Access to Snowden Trove." *Daily*

Beast, March 14, 2019. https://www.thedailybeast.com/the-intercept-shuts
-down-access-to-snowden-trove.

Taylor, Astra. "The Automation Charade." *Logic Magazine,* August 1, 2018.
https://logicmag.io/failure/the-automation-charade/.

———. *The People's Platform: Taking Back Power and Culture in the Digital Age.* London: Picador, 2015.

Taylor, Frederick Winslow. *The Principles of Scientific Management.* New York: Harper and Brothers, 1911.

Taylor, Jill Bolte. "The 2009 Time 100." *Time,* April 30, 2009. http://content
.time.com/time/specials/packages/article/0,28804,1894410_1893209_18
93475,00.html.

Theobald, Ulrich. "Liji." *Chinaknowledge.de,* July 24, 2010. http://www
.chinaknowledge.de/Literature/Classics/liji.html.

Thiel, Peter. "Good for Google, Bad for America." *New York Times,* August 1,
2019. https://www.nytimes.com/2019/08/01/opinion/peter-thiel-google
.html.

Thomas, David Hurst. *Skull Wars: Kennewick Man, Archaeology, and the Battle for Native American Identity.* New York: Basic Books, 2002.

Thompson, Edward P. "Time, Work-Discipline, and Industrial Capitalism." *Past and Present* 38 (1967): 56–97.

Tishkoff, Sarah A., and Kenneth K. Kidd. "Implications of Biogeography of Human Populations for 'Race' and Medicine." *Nature Genetics* 36, no. 11 (2004): S21–S27. https://doi.org/10.1038/ng1438.

Tockar, Anthony. "Riding with the Stars: Passenger Privacy in the NYC Taxicab Dataset." September 15, 2014. https://agkn.wordpress.com/2014/09/15
/riding-with-the-stars-passenger-privacy-in-the-nyc-taxicab-dataset/.

Tomkins, Silvan S. *Affect Imagery Consciousness: The Complete Edition.* New York: Springer, 2008.

Tomkins, Silvan S., and Robert McCarter. "What and Where Are the Primary Affects? Some Evidence for a Theory." *Perceptual and Motor Skills* 18, no. 1 (1964): 119–58. https://doi.org/10.2466/pms.1964.18.1.119.

Toscano, Marion E., and Elizabeth Maynard. "Understanding the Link: 'Homosexuality,' Gender Identity, and the *DSM.*" *Journal of LGBT Issues in Counseling* 8, no. 3 (2014): 248–63. https://doi.org/10.1080/15538605
.2014.897296.

Trainer, Ted. *Renewable Energy Cannot Sustain a Consumer Society.* Dordrecht, Netherlands: Springer, 2007.

"Transforming Intel's Supply Chain with Real-Time Analytics." Intel, September 2017. https://www.intel.com/content/dam/www/public/us/en/do
cuments/white-papers/transforming-supply-chain-with-real-time-analy
tics-whitepaper.pdf.

Tronchin, Lamberto. "The 'Phonurgia Nova' of Athanasius Kircher: The

Marvellous Sound World of 17th Century." Conference paper, 155th Meeting Acoustical Society of America, January 2008. https://doi.org/10.1121/1.2992053.

Tsukayama, Hayley. "Facebook Turns to Artificial Intelligence to Fight Hate and Misinformation in Myanmar." *Washington Post*, August 15, 2018. https://www.washingtonpost.com/technology/2018/08/16/facebook-turns-artificial-intelligence-fight-hate-misinformation-myanmar/.

Tucker, Patrick. "Refugee or Terrorist? IBM Thinks Its Software Has the Answer." Defense One, January 27, 2016. https://www.defenseone.com/technology/2016/01/refugee-or-terrorist-ibm-thinks-its-software-has-answer/125484/.

Tully, John. "A Victorian Ecological Disaster: Imperialism, the Telegraph, and Gutta-Percha." *Journal of World History* 20, no. 4 (2009): 559–79. https://doi.org/10.1353/jwh.0.0088.

Turing, A. M. "Computing Machinery and Intelligence." *Mind*, October 1, 1950, 433–60. https://doi.org/10.1093/mind/LIX.236.433.

Turner, Graham. "Is Global Collapse Imminent? An Updated Comparison of The Limits to Growth with Historical Data." Research Paper no. 4, Melbourne Sustainable Society Institute, University of Melbourne, August 2014.

Turner, H. W. "Contribution to the Geology of the Silver Peak Quadrangle, Nevada." *Bulletin of the Geological Society of America* 20, no. 1 (1909): 223–64.

Tuschling, Anna. "The Age of Affective Computing." In *Timing of Affect: Epistemologies, Aesthetics, Politics*, edited by Marie-Luise Angerer, Bernd Bösel, and Michaela Ott, 179–90. Zurich: Diaphanes, 2014.

Tversky, Amos, and Daniel Kahneman. "Judgment under Uncertainty: Heuristics and Biases." *Science* 185 (1974): 1124–31. https://doi.org/10.1126/science.185.4157.1124.

Ullman, Ellen. *Life in Code: A Personal History of Technology.* New York: MCD, 2017.

United Nations Conference on Trade and Development. *Review of Maritime Transport, 2017.* https://unctad.org/en/PublicationsLibrary/rmt2017_en.pdf.

U.S. Commercial Space Launch Competitiveness Act. Pub. L. No. 114–90, 114th Cong. (2015). https://www.congress.gov/114/plaws/publ90/PLAW-114publ90.pdf.

U.S. Congress. Senate Select Committee on Intelligence Activities. *Covert Action in Chile, 1963–1973.* Staff Report, December 18, 1975. https://www.archives.gov/files/declassification/iscap/pdf/2010-009-doc17.pdf.

U.S. Energy Information Administration, "What Is U.S. Electricity Genera-

tion by Energy Source?" https://www.eia.gov/tools/faqs/faq.php?id=427
&t=21.

"Use of the 'Not Releasable to Foreign Nationals' (NOFORN) Caveat on Department of Defense (DoD) Information." U.S. Department of Defense,
May 17, 2005. https://fas.org/sgp/othergov/dod/noforn051705.pdf.

"UTKFace—Aicip." http://aicip.eecs.utk.edu/wiki/UTKFace.

Vidal, John. "Health Risks of Shipping Pollution Have Been 'Underesti-
mated.'" *Guardian,* April 9, 2009. https://www.theguardian.com/environ
ment/2009/apr/09/shipping-pollution.

"Vigilant Solutions." NCPA. http://www.ncpa.us/Vendors/Vigilant%20
Solutions.

Vincent, James. "AI 'Emotion Recognition' Can't Be Trusted.'" *The Verge,*
July 25, 2019. https://www.theverge.com/2019/7/25/8929793/emotion-rec
ognition-analysis-ai-machine-learning-facial-expression-review.

———. "Drones Taught to Spot Violent Behavior in Crowds Using AI."
The Verge, June 6, 2018. https://www.theverge.com/2018/6/6/17433482/ai
-automated-surveillance-drones-spot-violent-behavior-crowds.

Vollmann, William T. "Invisible and Insidious." *Harper's Magazine,* March
2015. https://harpers.org/archive/2015/03/invisible-and-insidious/.

von Neumann, John. *The Computer and the Brain.* New Haven: Yale University, 1958.

Wade, Lizzie. "Tesla's Electric Cars Aren't as Green as You Might Think."
Wired, March 31, 2016. https://www.wired.com/2016/03/teslas-electric
-cars-might-not-green-think/.

Wajcman, Judy. "How Silicon Valley Sets Time." *New Media and Society* 21,
no. 6 (2019): 1272–89. https://doi.org/10.1177/1461444818820073.

———. *Pressed for Time: The Acceleration of Life in Digital Capitalism.* Chicago: University of Chicago Press, 2015.

Wakabayashi, Daisuke. "Google's Shadow Work Force: Temps Who Outnumber Full-Time Employees." *New York Times,* May 28, 2019. https://
www.nytimes.com/2019/05/28/technology/google-temp-workers.html.

Wald, Ellen. "Tesla Is a Battery Business, Not a Car Business." *Forbes,* April
15, 2017. https://www.forbes.com/sites/ellenrwald/2017/04/15/tesla-is-a
-battery-business-not-a-car-business/.

Waldman, Peter, Lizette Chapman, and Jordan Robertson. "Palantir Knows
Everything about You." *Bloomberg,* April 19, 2018. https://www.bloom
berg.com/features/2018-palantir-peter-thiel/.

Wang, Yilun, and Michal Kosinski. "Deep Neural Networks *Are* More Accurate Than Humans at Detecting Sexual Orientation from Facial Images."
Journal of Personality and Social Psychology 114, no. 2 (2018): 246–57.
https://doi.org/10.1037/pspa0000098.

"The War against Immigrants: Trump's Tech Tools Powered by Palantir."
Mijente, August 2019. https://mijente.net/wp-content/uploads/2019/08
/Mijente-The-War-Against-Immigrants_-Trumps-Tech-Tools-Powered
-by-Palantir_.pdf.

Ward, Bob. *Dr. Space: The Life of Wernher von Braun.* Annapolis, Md.: Naval
Institute Press, 2009.

Weigel, Moira. "Palantir goes to the Frankfurt School." *boundary 2* (blog),
July 10, 2020. https://www.boundary2.org/2020/07/moira-weigel-palan
tir-goes-to-the-frankfurt-school/.

Weinberger, Sharon. "Airport Security: Intent to Deceive?" *Nature* 465
(2010): 412–15. https://doi.org/10.1038/465412a.

Weizenbaum, Joseph. *Computer Power and Human Reason: From Judgment
to Calculation.* San Francisco: W. H. Freeman, 1976.

———. "On the Impact of the Computer on Society: How Does One Insult
a Machine?" *Science,* n.s., 176 (1972): 609–14.

Welch, Chris. "Elon Musk: First Humans Who Journey to Mars Must 'Be
Prepared to Die.'" *The Verge,* September 27, 2016. https://www.theverge
.com/2016/9/27/13080836/elon-musk-spacex-mars-mission-death-risk.

Werrett, Simon. "Potemkin and the Panopticon: Samuel Bentham and the
Architecture of Absolutism in Eighteenth Century Russia." *Journal of
Bentham Studies* 2 (1999). https://doi.org/10.14324/111.2045-757X.010.

West, Cornel. "A Genealogy of Modern Racism." In *Race Critical Theories:
Text and Context,* edited by Philomena Essed and David Theo Goldberg,
90–112. Malden, Mass.: Blackwell, 2002.

West, Sarah Myers. "Redistribution and Rekognition: A Feminist Critique
of Fairness." *Catalyst: Feminism, Theory, and Technoscience* (forthcoming, 2020).

West, Sarah Myers, Meredith Whittaker, and Kate Crawford. "Discriminating Systems: Gender, Race, and Power in AI." AI Now Institute, April
2019. https://ainowinstitute.org/discriminatingsystems.pdf.

Whittaker, Meredith, et al. *AI Now Report 2018.* AI Now Institute, December
2018. https://ainowinstitute.org/AI_Now_2018_Report.pdf.

———. "Disability, Bias, and AI." AI Now Institute, November 2019. https://
ainowinstitute.org/disabilitybiasai-2019.pdf.

"Why Asteroids." Planetary Resources. https://www.planetaryresources.com
/why-asteroids/.

Wilson, Mark. "Amazon and Target Race to Revolutionize the Cardboard
Shipping Box." *Fast Company,* May 6, 2019. https://www.fastcompany
.com/90342864/rethinking-the-cardboard-box-has-never-been-more
-important-just-ask-amazon-and-target.

Wilson, Megan R. "Top Lobbying Victories of 2015." The Hill, December

16, 2015. https://thehill.com/business-a-lobbying/business-a-lobbying/263354-lobbying-victories-of-2015.

Winston, Ali, and Ingrid Burrington. "A Pioneer in Predictive Policing Is Starting a Troubling New Project." *The Verge* (blog), April 26, 2018. https://www.theverge.com/2018/4/26/17285058/predictive-policing-predpol-pentagon-ai-racial-bias.

Winner, Langdon. *The Whale and the Reactor: A Search for Limits in an Age of High Technology.* Chicago: University of Chicago Press, 2001.

Wood, Bryan. "What Is Happening with the Uighurs in China?" PBS News-Hour. https://www.pbs.org/newshour/features/uighurs/.

Wood III, Pat, William L. Massey, and Nora Mead Brownell. "FERC Order Directing Release of Information." Federal Energy Regulatory Commission, March 21, 2003. https://www.caiso.com/Documents/FERCOrderDirectingRelease-InformationinDocketNos_PA02-2-000_etal__Manipulation-ElectricandGasPrices_.pdf.

Wu, Xiaolin, and Xi Zhang. "Automated Inference on Criminality Using Face Images." *arXiv:1611.04135v1 [cs.CV],* November 13, 2016. https://arxiv.org/abs/1611.04135v1.

Yahoo! "Datasets." https://webscope.sandbox.yahoo.com/catalog.php?datatype=i&did=67&guccounter=1.

Yang, Kaiyu, et al. "Towards Fairer Datasets: Filtering and Balancing the Distribution of the People Subtree in the ImageNet Hierarchy." In *FAT* '20: Proceedings of the 2020 Conference on Fairness, Accountability, and Transparency,* 547–558. New York: ACM Press, 2020. https://dl.acm.org/doi/proceedings/10.1145/3351095.

"YFCC100M Core Dataset." Multimedia Commons Initiative, December 4, 2015. https://multimediacommons.wordpress.com/yfcc100m-core-dataset/.

Yuan, Li. "How Cheap Labor Drives China's A.I. Ambitions." *New York Times,* November 25, 2018. https://www.nytimes.com/2018/11/25/business/china-artificial-intelligence-labeling.html.

Zhang, Zhimeng, et al. "Multi-Target, Multi-Camera Tracking by Hierarchical Clustering: Recent Progress on DukeMTMC Project." *arXiv:1712.09531 [cs.CV],* December 27, 2017. https://arxiv.org/abs/1712.09531.

Zuboff, Shoshana. *The Age of Surveillance Capitalism: The Fight for a Human Future at the New Frontier of Power.* New York: PublicAffairs, 2019.

———. "Big Other: Surveillance Capitalism and the Prospects of an Information Civilization." *Journal of Information Technology* 30, no. 1 (2015): 75–89. https://doi.org/10.1057/jit.2015.5.

Index

Figures and notes are indicated by f and n following the page number.

Microsoft: affect recognition used
by, 155; carbon footprint of,
43–44; carbon footprint of data
centers, 41; facial recognition
and, 253n46; MS-Celeb dataset, 110; Project Maven and, 191,
263n36
Microworkers, 64
mineral extraction: AI's need for,
32–36; conflict linked to, 34–35;
myth of clean tech, 41–46;
supply chains, 34–35; true costs
of, 26, 36–41. *See also* extractive
industries
Minsky, Marvin, 5, 6, 245n10
Mirzoeff, Nicholas: *The Right to
Look*, 62
MIT: affect recognition research,
154, 169; "Management and the
Computer of the Future" lecture
series, 5–6; Media Lab, 169
Moffett Federal Airfield, 23
Mongolia: lithium mining in, 33;
rare earth mineral extraction
in, 36–37
Morozov, Evgeny, 110
Morton, Samuel, 123–26, 127
MS-Celeb dataset, 110, 220
mug shot databases, 89–92, 93
Multiple Encounter Dataset, 89–
92
Mumford, Lewis, 33, 48
Muse, Abdi, 84–85
Musk, Elon, 29, 231, 233
myth of clean tech, 41–46

National Health Service (UK), 120
National Institute of Standards and
Technology (NIST), 89–92, 93,
220, 252n3
National Science Foundation, 109

National Security Agency (NSA):
data center, 44–45; drone strike
targeting and, 203; Palantir and,
194; Project Maven and, 189–90;
state power and, 181–83, 183f,
185, 188, 209, 222; surveillance
by, 194–95
natural language processing (NLP)
model, 42–43
necropolitics, 185
Negri, Antonio, 217
Neighbors (app), 201, 202
Neilson, Brett, 247n21
Nelson, Alondra, 12, 20
neodymium, 33
Network Time Protocol (NTP),
77–78
Neville-Neil, George V., 265n6
Newell, Allen, 6
Newton, Isaac, 79
New York City Taxi and Limousine
Commission, 110–11
New York Taxi Workers Alliance,
86
New York Times: on Clever Hans, 1;
on Ekman, 170
Nextdoor (app), 201
Next Generation Identification
(NGI), 92
Nietzsche, Friedrich, 222
Nilsson, Nils, 111–12
NIST. *See* National Institute of
Standards and Technology
NLP (natural language processing)
model, 42–43
Noble, Safiya Umoja, 13, 128
Norvig, Peter, 7, 11
NSA. *See* National Security
Agency
NTP (Network Time Protocol),
77–78